MEDIA HOT AND COLD

ELEMENTS *A series edited
by Stacy Alaimo and Nicole Starosielski*

NICOLE   STARO   SIELSKI

MEDIA

HOT

&

COLD

DUKE UNIVERSITY PRESS   Durham and London   2021

© 2021 Duke University Press
All rights reserved
Printed and bound by CPI Group (UK) Ltd, Croydon, CR0 4YY
Cover design by Aimee C. Harrison
Typeset in Chaparral Pro and Trade Gothic LT Std by Westchester
Publishing Services

Library of Congress Cataloging-in-Publication Data
Names: Starosielski, Nicole, [date] author.
Title: Media hot and cold / Nicole Starosielski.
Other titles: Elements (Duke University Press)
Description: Durham : Duke University Press, 2021. | Series: Elements |
Includes bibliographical references and index.
Identifiers: LCCN 2021012474 (print)
LCCN 2021012475 (ebook)
ISBN 9781478013617 (hardcover)
ISBN 9781478014546 (paperback)
ISBN 9781478021841 (ebook)
Subjects: LCSH: Body temperature—Regulation—Social aspects. |
Temperature sense—Social aspects. | Global temperature changes. |
Human beings—Effect of climate on. | BISAC: SOCIAL SCIENCE / Media
Studies | SCIENCE / General
Classification: LCC QP135.S744 2021 (print) | LCC QP135 (ebook) |
DDC 612/.01426—dc23
LC record available at https://lccn.loc.gov/2021012474
LC ebook record available at https://lccn.loc.gov/2021012475

Cover art: (1) Craig Nisbet, *TV Static*. Source: Filter Forge. (2) Kevin
Doncaster, *Infrared*, licensed under CC BY 2.0. Source: Flickr. (3) Tyler
Nienhouse, photograph of red fibers, licensed under CC BY 2.0. Source:
Flickr. (4) Image taken with infrared camera. Each color represents differ-
ent temperatures, as is shown on spectrum scale on right side of image.
Source: iStock. (5) The infrared sensor from the array of sensors on the
front at the top of a BlackBerry Leap, licensed under CC BY 4.0. Courtesy
of D-Kuru. Source: Wikimedia Commons. (6) Sheila Sund, *Sixty-Eight
Degrees*, licensed under CC BY 2.0. Source: Flickr. (7) All other images in
public domain or licensed under Creative Commons.

for and with Jamie

Acknowledgments   ix

Preface: Of Temperature   xiii

Introduction: Media Hot and Cold   1

PART I

1   THERMOSTAT: THE THERMAL SUBJECTS OF BROADCAST TEMPERATURE   31

2   COLDSPLOITATION: THE THERMAL ATTRACTIONS OF COOL AIR   72

3   SWEATBOX: THE THERMAL VIOLENCE OF WEAPONIZED HEAT   109

PART II

4   HEAT RAY: THE THERMAL CIRCUITS OF RADIANT MEDIA   135

5   INFRARED CAMERA: THE THERMAL VISION OF HEAT IMAGES   166

6   COMPUTER: THE COLDWARD COURSE OF MEDIA   191

Conclusion: Media after the Melt   219

Notes   225

Bibliography   255

Index   273

Writing this book has been an absorptive process. Everyone I speak with has something to share about temperature: observations, sensations, memories, chance encounters, and objects. These exchanges have found their way into the following pages, and this book has benefited enormously from them. I want to thank Shannon Mattern for being the most generous reader I could imagine; Stefan Helmreich for attuning me to waves; Fernando Domínguez Rubio for the joint investigations into thermal ecologies; Kyle Devine and Jacob Smith for sound feedback; Regina Longo for archival wisdom; Natasha Schüll for helping me to track heat with a borrowed Embr Wave; Jeff Scheible for laying everything on ice; Susan Zieger for thinking outside the box; Tracy for the phase state changes; Rahul Mukherjee for radiant awareness; Erica Robles-Anderson for ambient and architectural awareness; Julia Velkova and Yarden Katz for meditations on ecological computing; Daniel Barber for broadening my sense of climate; John Durham Peters for asking, among many other things, "What about warm?"; and Alex Galloway for carefully reading the manuscript in its final iteration. I would like to thank the many colleagues who directed me to cases of heat and cold, read pieces of the manuscript in progress, and shepherded sections of *Media Hot and Cold* into publication—especially Brooke Belisle, Elena Beregow, Nathan Ensmenger, Ilana Gershon, Anna McCarthy, Dylan Mulvin, Jussi Parikka, Rachel Plotnick, James Schwoch, and Jonathan Sterne.

I composed the bulk of this manuscript while editing the Elements series at Duke University Press, and I am grateful to my elemental coconspirators— to Darin Barney for the political grounding; Yuriko Furuhata and Marina Peterson for the atmospheric dispersions; Heather Davis for complicating what matters; Anne Pasek for the never-ending fixes; Eva Hayward and Melody Jue for submersion and depth; Hi'ilei Julia Kawehipuaakahaopulani Hobart, Rafico Ruiz, and Jen Rose Smith for destabilizing thermal terrain; and Chris Russill and Liam Cole Young for earthly recompositions. This book is deeply entangled with their imaginative writings on water, carbon, ice, earth, salt, air, oil, and plastic. I have been incredibly fortunate to have the insight and support of Courtney Berger, whose capacious editorial work and expansive vision paved the way for the series and this book, and Stacy Alaimo, whose questions over the past several years reverberate through the following pages. The editoral team at Duke University Press has been wonderful, and I would like to thank Aimee C. Harrison and Lisl Hampton.

The first inklings of this book took shape at the University of California, Santa Barbara, and I remain thankful for the ongoing inspiration of Maria Corrigan, Bishnu Ghosh, Jen Holt, Joshua Neves, Rita Raley, Bhaskar Sarkar, and Janet Walker. As always, I have immense gratitude for Lisa Parks, whose work to expand the landscape of media studies set the foundation for this project, without which I couldn't even have imagined this book. At New York University, I am especially grateful for the research support of Ian Alexander, Rodrigo Ferreira, Colette Perold, Victoria Pihl Sorensen, and Annie Tressler—their work has been invaluable. The members of the INFRA and fiber research groups were wonderful interlocutors during the book's early formulations, and I would like to thank the graduate students who explored temperature alongside me, including Neta Alexander, Mei Ling Chua, Sam Kellogg, Harris Kornstein, Leonard Santos, Rory Solomon, and Meg Wiessner. I am grateful to my colleagues in the Department of Media, Culture, and Communication for the many conversations about heat and cold over the past years, especially Isra Ali, Arjun Appadurai, Finn Brunton, Lily Chumley, Lisa Gitelman, Radha Hegde, Ben Kafka, Mara Mills, Nick Mirzoeff, Kelli Moore, Sue Murray, Juan Pinon, Helga Tawil-Souri, and Angela Xiao Wu.

I had the opportunity to present iterations of this book at several institutions, where I received wonderful feedback from colleagues. I would like to thank the organizers and audiences of the Hardwired Temporalities conference; the Solarity: After Oil School; the Materials for Interaction conference at Indiana University, Bloomington; as well as the Department of Modern Culture and Media at Brown University; the Department of Media, Journalism and Film at Miami University (Ohio); the English Department at the University of Pennsylvania; the Department of Communication at the University of California, San Diego; and the Film and Media Studies Program at Yale University; as well as the Digital Assembly at the University of Florida; Columbia University's Seminar on the Theory and History of Media; and the GIDEST seminar at the New School. Portions of this manuscript appeared as "The Materiality of Media Heat," *International Journal of Communication* 8 (2014): 2504–8; "Thermocultures of Geological Media," *Cultural Politics* 12, no. 3 (2016): 293–309; "Thermal Violence: Heat Rays, Sweatboxes and the Politics of Exposure," *Culture Machine* 17 (2019): 1–27; and "Thermal Vision," *Journal of Visual Culture* 18, no. 2 (2019): 1–23.

Media Hot and Cold was written for and with Jamie Skye Bianco, who built the fires, checked the thermometers, and composed a thermal landscape for all of us to inhabit.

I wrote this book in the heat and the cold, in fields and under plastic hoops, in barns and chicken coops, as I built a farm in the interstices of another life. Farming is a practice of temperature. It is the long, slow waiting for the last night that dips below freezing. It is the rush that happens when the heat arrives and everyone runs out to the fields. It is a constant, ever-present attention to slight changes in heat. It is the sensation of standing in frigid water. It is the blast of cold air in walk-in refrigerators. It is watching produce wilt because of radiant pavement. It is moments of thermal recognition: Seedlings baked under too much plastic. An entire crop wiped out by an unexpected frost. It is the warmth of mammals born on winter nights. It is the heat that lingers in an animal's body after it dies. It is the blanket, the hutch, the straw, the hay, the tarp, the paper, the plastic, and a million other things overlaid and removed in order to maintain radiant life.

I wrote these words in short bursts between these practices of temperature, often when it was too cold or too hot to be outside. Some were written on a cell phone in the field, others scrawled on paper encrusted with dirt, shit, and blood. Theory always bears the traces of the author's environment, as Melody Jue so eloquently explains in *Wild Blue Media*. The theory of heat and cold in the following pages does not carry the sensory residue of an air-conditioned office. It emerges from a radically variable thermal world, with exposures ranging from −34°C/−30°F to over 37°C/100°F. Although this book is not about farming, it brings with it the traces of agricultural labor.

Media Hot and Cold is an experiment in writing and inhabitation.[1] In the field I wondered: Can words actually transform thermoception, the sense of temperature? If so, what can they evoke and elicit? How is temperature patterned not only for people insulated in air-conditioned offices but also for those who remain immersed and vulnerable in their environments? Can thermal media, whether films or fans, offer other ways of being in temperature? While these questions crystallized for me in the field, I found the answers in the thermal accounts of others. The following pages are an archive of thermal experiences: women subjected to chilly workplaces; cinema exhibitors exploiting the cold; people punished in sweatboxes; doctors producing artificial fevers; and conservationists using thermal imaging to manage wildlife, among many others. Looking for heat and cold, I quickly found that temperature is neither a neutral nor a natural environmental register; instead, it's thoroughly entangled with cultural practices and media technologies.

On the farm, I found an analogue for book writing while watching chicks gather around a heat lamp. After years of using digital thermometers to manage the birds, we have come to rely simply on sight. When the birds are clustered at the edge of the brooder, it is too hot. When the birds are evenly distributed, the temperature is just right. I would only later discover that the first application of electrical thermostats was to standardize incubation temperatures on chicken farms.[2] Farming attunes one's body to differentiated thermal zones. It cultivates synesthesia, ways to thermosense through hearing and vision. At its core, this is also the project of *Media Hot and Cold*. I hope that this book will cultivate a synesthetic attunement, make social activity reveal its invisible thermal contexts, and orient readers to thermal fields. In short, this is an experiment in temperature manipulation, intended to work as farming does: on and through the body.

The most influential messages of the twenty-first century will be sent not through words and images but through heat and cold. People will be turned around at borders not only by signs but also by air conditioners that withdraw feeling from their fingertips. Thermal cameras will scan bodies for fever and sickness. Militaries will shoot heat rays at target populations. Overheated prisons will tell some people that they are less than human. Other people will learn that climate change is their own problem to deal with, an individual discomfort to be overcome with a digital app. Temperature differences, read by satellites, will indicate sites for mineral extraction more effectively than they predict the coming weather. The internet's content will be transcoded into infrared waves and sent through fiber-optic cables. As temperature volatility increases, thermal media—thermostats, heating and cooling systems, architectures, and infrared cameras—will carry the promise of stabilization and social control. But their deployment and effects will be uneven. They will mitigate the impact of climate change for some and intensify its harm for others.

The most powerful media organizations of the twenty-first century will be thermal. The circulation of images, sounds, videos, and texts will depend on a massive regime of heating and cooling. Data and networks, like the people they connect, will be ever more fragile. Too hot or too cold, and the platforms will collapse. Digital infrastructures—data centers, network exchanges, and fiber-optic cables—will drain the planet's energy in order to create a stable thermal environment—not for people but for information. Meanwhile, designers and operators of thermal media will make decisions about how information will travel, how quickly the planet will warm, and who will survive. They will establish standards for acceptable temperatures. They will modulate and regulate bodily exposures. Their products will cultivate thermal expectations. In the process, they will expand zones of thermal privilege and thermal harm. Bodies will be subjugated and populations will be organized not only through imprisonment and torture, surveillance systems and databases, but also through a vast regime of temperature management. Technologies from wearable air conditioners to water cannons will be the instruments of thermopower.

Heat and cold are everywhere a means of communication, manipulation, and subjugation, but they often appear simply as natural phenomena. There is widespread faith in thermal objectivity, the sense that temperature is independent of both culture and perception. Even though heat is used to differentiate people and places, it is often described as universal and democratizing. After all, heat travels in and through everything. It affects the metabolisms of all bodies and influences the operations of all matter. Everything is, fundamentally, entangled with temperature. Thermodynamics, the science of energetic conversion and thermal transformation, has been powerful in part because of this generality. One textbook boasts that thermodynamics "can be applied to any discipline, technology, application, or process."[1] Thanks to thermal objectivity, heat is not only a subject of scientific study but also a metric. Temperature offers a universal language—degrees Celsius and Fahrenheit (and Kelvin)—that can describe engines, people, and climate change alike. In turn, temperature's meanings can be taken for granted by scientists, architects, meteorologists, environmentalists, and the public. Thermal objectivity has immense political potential. A global anticipation of a 2°C/3.6°F increase in temperature sparks political action on climate change. Governments define thermal rights—the right to heat and, in some instances, the right to air-conditioning.

Although thermal objectivity grounds climate science and the everyday politics of thermal exposure, it often masks the operations of thermopower. Descriptions of heat waves bind people in a shared perception of their surroundings, but they often obscure how thermal media redistribute heat effects.[2] The common sense of "hot" and "cold" obscures vernacular thermal perceptions. A focus on the thermometer obscures how body temperature is managed through color, screens, and tactile interfaces. *Media Hot and Cold* investigates the many meanings and messages of heat and cold, the ways that people manipulate heat signals, and, in particular, the technologies that relay thermal communications. Instead of situating temperature as a neutral backdrop for social life, this book tracks how heat and cold are mediated by expansive thermal regimes.

Attuning to heat has never been more critical than in the current moment, which is marked by the intensification of climate change as well as atmospheric communication, control, and contagion. The argument of this book is that these forces are catalyzing a shift in the operation of thermopower. They elevate thermopower—the enactment of social and political power through thermal manipulation—to a pervasive means of biological, social, and environmental control. At the same time, temperature is being tethered to digital systems. Heat and cold are now experienced in ways that mirror shifts across the media landscape: they are increasingly personalized, marketed as a means of networked affective consumption, and digitally modulated as part of weaponized sensory environments. Tracking the history of this transition—from the mass thermal communications of the twentieth century to the digital thermal technologies of the twenty-first century—*Media Hot and Cold* traces how media set baselines for our thermal futures. Its conclusion runs counter to arguments that locate digital and networked technologies as a means of stabilization in the midst of climate change. Instead, I argue that local and autonomous forms of thermal communication are essential to truly mitigate climate effects.

ATTUNING TO HEAT

To attune readers to temperature's political operations, *Media Hot and Cold* begins with a premise radically different from thermal objectivity. In this book, temperature is not a property of bounded entities—the measure of a place, a body, or a building. Rather, temperature is one way of describing the exchange of heat, a process in which everything participates. All matter emits heat: the skin of a human body, the asphalt of a road, the hard plastic of a container. Plant leaves emit heat even as they stretch up toward the sun. A sweatshirt radiates heat even as it also traps it. All media emit heat. Phones transfer heat to hands. Broadcast towers, cellular antennae, and satellites emit thermal radiation alongside radio waves. Heat moves through these pages, through bodies, clothes, architecture, and atmospheres. We (and this "we" extends to the nonhuman, the animal and the geophysical) are constantly communicating by generating heat signals. These are often affective emissions, relayed without intentionality and registered without cognition.

At the same time, we—bodies, objects, and infrastructures—are also heat receivers, affected by heat and cold. We are caught in the midst of emissions, both waves and electrons, generated by all of the promiscuous heat producers around us. One goat huddles with others on a cold night, receiving their warmth and, in turn, contributing its own body heat. The herd is created in the exchange of heat, not solely through vision or sound. Observing humans rather than ruminants, some argue that the drive toward heat is a drive toward the mother, a drive to return to the womb and to a state of shared bodily warmth. Others point out that thermal exchanges can reinforce species, familial, and sexual connections. Warmth, as Lauren Berlant writes, "is an atmosphere that allows life and death to be in the same place as what's potentially unbearable in love every minute."[3] Feeling the heat of others can also be a queer process. Packed together on a subway car, in an auditorium, on a dance floor, or in the streets, people receive heat signals from other bodies. They slide in warm sweat and brush up against clammy skin. This heat might be sensed as a crowd, a community, or a collective.[4] Or it might be perceived as a threat, oppression, or a means of controlling their bodies from afar. Social distancing mutes thermal transmissions, extracting some people from interpersonal heat networks. People use heat and its absence to navigate the world. Mosquitoes gravitate toward warm blood, pit vipers sense prey from a distance, and drone operators view the landscape through thermal cameras. Bodies are oriented by thermoception.

Temperature also stimulates responses without any perception at all. A building does not perceive the fluctuation between hot and cold, but it is sensitive to changes in the weather. On a hot day, its large windows heat up. The glass expands, first in the center of the window and later next to the frame. As night falls, the temperature drops quickly. The difference between hot and cold is stressful. The glass cracks—this is called a thermal fracture. Fluctuating temperatures have generated fragility, and the window breaks. Bodies inside the building, now subject to a draft, lose heat at their edges; fingers and toes are the first to cool. Changes in temperature alter a building's potential to shelter people. They also alter how long a virus remains viable. And they shape a body's capacity to survive. Even in the absence of the sense of thermoception, thermosensitivities cascade. Temperature catalyzes waves of transmission. Heat signals come to matter in a multitude of ways that escape conscious perception.

In other words, the transmission of heat and cold is a form of communication, even if in many instances what they relay—any "message" sent, received, interpreted, and reacted to—is diffuse and distributed. But despite their ubiquity, thermal communications have rarely been a part of understanding media.[5] Instead, media studies has historically focused on texts and images, turning only more recently to sonic and haptic technologies. Thermoception is marginal even in the environmental humanities and social sciences, research areas committed to understanding the cultural dimensions of environmental phenomena. In the expansive and dynamic field of scholarship documenting environmental news, literature, and media, heat and cold are rarely described as communicative forms, even though they have been used to enact violence, sell products, and send information. In order to understand images, scholars analyze practices of light and visual culture. Sound studies has emerged in order to conceptualize aural transmissions. In *Media Hot and Cold*, I argue that in order to understand temperature, scholars must attend to practices of heat and to thermal cultures. This book offers a sensory paradigm that can make practices of heat and cold thermoceptible.[6]

Encounters with the field of heat exchanges, this book shows, are not neutral interactions with an external environment. They are always shaped by thermocultures, or thermal cultures: the cultural processes of thermal modulation and exposure that pattern sensitivities to heat and cold, construct normative ways of making sense of thermal stimuli, and set expectations for temperature. One aim of this book is to evoke a multitude of thermocultures. It might not be surprising that many thermocultures mirror other cultural formations. For example, just as the visual blankness of an Arctic landscape is deployed as part of colonial projects, its temperature—which to bodies of white settlers might feel "foreign"—is enrolled in projects of dispossession. This is thermal colonization: the use of a sense of temperature to justify colonial expansion and inhabitation. Or consider how the central cooling of an office building, set to standards for cisgender men, signifies a failure to accommodate women.[7] Corporations then deploy personalized systems that enable people to regulate their own temperature. This is thermal neoliberalism, which off-loads thermal responsibility onto the individual.

As is true for visual, aural, and textual cultures, there are dominant and marginalized thermocultures. In many Western contexts, hot and cold are sensed as antithetical, as oppositional extremes. For Hawaiians, as Hiʻilei Julia Kawehipuaakahaopulani Hobart documents in a groundbreaking study of ice, the sensation of heat is historically closer to the sense of cold than either is to a "neutral" temperature.[8] And although global weather reporting assumes its viewers share a common understanding of what temperature feels hot or cold, Mary Douglas's anthropological account of people in the Congo testifies to the local specificity of thermoception: although they experience the same weather, two tribes locate their hot and cold seasons at opposite points of the calendar.[9] The emission and transmission of thermal affect is not only a physical or physiological process. Like cultures of viewing and listening, thermocultures emerge out of existing practices and take shape in existing structures of meaning.

Thermocultures, like visual and sonic cultures, are deeply technological. A vast infrastructure of heating and cooling conditions bodies, objects, and materials. The air-conditioning of a retail store during a hot summer is engineered to draw people in and produce a space for consumption. The air-conditioning of concrete rooms where data is housed facilitates the circulation of digital media. Networked and intelligent thermostats learn personal preferences, producing an ever more controlled, sensitive, and seemingly safe environment for subjects. Infrared cameras target bodies at international airports, looking for signs of fever and threat to domestic populations. Thermometers produce "objective" and "standardized" bodies for medical practice.[10] Thermal media are the material and socially realized forms that communicate temperature, enabling heat and cold to be transmitted and received.[11] Thermal media relay thermal affects, craft thermal subjectivities, and calibrate thermoception. They communicate to people and modulate the thermal experiences of nonhumans. *Media Hot and Cold* describes the operations of thermal media, ranging from infrared cameras and air-conditioning units to heat ray guns and coldsploitation cinema. These forms of communication function as many other media do: They standardize and scale transmissions. They establish pathways for social interaction. They reorganize the mediating capacities of bodies, materials, and environments. They facilitate the encoding and decoding of meaning. They instill modes of sensation. In the process, they also emerge in relation to the landscape of existing media: as broadcast, narrowcast, or personalized forms.

Despite the apparent universality of heat, its effects and affects are not evenly distributed. *Media Hot and Cold* tracks these inequities. It shows how, through specific thermal media, thermoceptive regimes—the dominant material and ideological formations that set normative responses to heat and cold—are embedded into architectures and pierce through the skin. Take the case of a public housing complex in Chicago where residents for decades had been accustomed to the extraordinary heating of their buildings, colloquially referred to as "Project Heat." As Catherine Fennell documents in a sensory ethnography of the building, when the projects were sold, residents were newly able to control the heat and would regularly set it at a high temperature.[12] Yet they still had to wear several layers of clothing to keep warm. Project Heat had been programmed into their bodies. Their sense of environment had been calibrated to an unsustainable norm, both in relation to the economic realities of the newly privatized infrastructure and the extractive energy regimes it depended on. Communication by heat and the technologies that reconfigure it are always political. Thermoceptive regimes enact racialized, colonial, gendered, and sexualized forms of power, and they instill these forms of power as embodied sensation.

This book is not about how nature seeks to undo gradients (a core principle of thermodynamics) but rather dwells in the many forms of thermal difference. It describes how heating and cooling infrastructures—and forms of thermal privilege—are made available for some people and not for others. Architectures subject some inhabitants, and not others, to thermal violence. The freeze is sold as an attraction to some viewers, and not others. Heat maps illuminate some bodies, and not others. Smelters produce industrial materials for customers while generating harsh thermal conditions for their operators. Even the thermometer, as Deanna Day demonstrates, did not always appear to be a neutral carrier of thermal objectivity: it threatened physicians' authority, evoked female sexual desire, and was taken up as a gendered tool for reproductive labor.[13] Because it differentially activates (and thus sorts) bodies, thermal mediation helps to produce gender, race, class, ethnicity, and other forms of social difference. Although heat exchanges may be universal, they are certainly not uniform. This book is a story of thermal media's role in their uneven distribution, a study of the thermal fractures that structure contemporary life.

The word *thermopower* describes the mechanisms of this uneven distribution—the ways that temperature management defines subjects, produces objects, and locates both in grids of social and political organization. As thermocultures expose people to particular thermal worlds and normalize thermoceptive regimes in which heat and cold make sense, thermopower operates as a form of biopower, a means of administering and regulating life. In the process, thermopolitics also operates as a form of necropolitics, what Achille Mbembe describes as the capacity to expose to death.[14] But thermopower's subjects are not only people. Thermal media construct and dismantle environments. They are integral to the creation of almost all industrial and postindustrial technologies. Their communicative forms extend far beyond the human. Indeed, thermal media are often used to distinguish between life and nonlife, operating as vectors of what Elizabeth Povinelli calls geontopower. In short, thermopower extends through these many other forms of political power. It is defined neither by its subjects nor by its particular operations. Like energopower, the operation of political power through fuel and electricity, thermopower is defined by its media: heat and cold.[15] It is the enactment of power across all of these domains— the biological, the geological, and the energetic—through the mediation of temperature.

Media Hot and Cold attunes readers to thermocultures, documents the work of thermal media, and tracks their imbrication with regimes of thermopower in order to congeal an emerging field of inquiry. What I call critical temperature studies approaches thermal objectivity as a historically specific system of knowledge based in Western science. Counter to thermodynamics and other universalizing lineages, critical temperature studies foregrounds the investigation of the material, political, cultural, and representational dimensions of heat, warmth, and cold. It is committed to addressing how temperature materializes in divergent ways across geographies and history, and to documenting forms of thermal violence and colonization. It locates thermoception alongside sight and hearing.

Critical temperature studies is already well developed in anthropology, energy studies, geography, architecture, and studies of affect and the senses, among other fields.[16] Architecture, as Boon Lay Ong defines it, "is the orchestration of heat through energy, climate, and habitation."[17] Buildings function as "climate mediators," Daniel Barber shows, not only through their mechanical systems but in the design of their facades.[18] Historians have chronicled the cultural dynamics of heating, cooling, and thermometric technologies, as well as the "invention" of temperature itself.[19] Media scholars have documented the communicative work of fire and matter's shifting phase states, and the difficulty of representing heat waves in journalism.[20] The critical study of cryopolitics reveals how life is managed through refrigerating technologies, and the "cold humanities" investigates the experience of cold on a warming planet.[21] Temperature, Fernando Domínguez Rubio points out, is a crucial part of the ecologies through which "'social' and 'cultural' worlds become possible."[22] And as geographer Mike Hulme describes, the very idea of climate functions to stabilize cultural relationships between people and their weather.[23] Ethnographers have observed cultures of thermoception and the "temperature work" that scaffolds them.[24] In radically different contexts, Eva Horn and Bharat Jayram Venkat have called for an anthropology of climate and of heat that can account for the many connections between thermal environments and culture.[25]

These studies are the foundation of *Media Hot and Cold*. Such works often remain within disciplinary confines, heat or cold specific, tied to geographic areas or historical periods, oriented toward the natural or built environment, and anchored in particular forms of human sensation or nonhuman agency. *Media Hot and Cold* locates these as part of a shared inquiry into temperature's social and cultural dimensions. Like visual culture and sound studies, work in critical temperature studies melds research across disciplines, geographies, and historical periods. As recent works in the environmental humanities, new materialisms, and sensory studies suggest, such research extends beyond the human sensation of temperature to its non-anthropomorphic possibilities.[26] To this field of inquiry, *Media Hot and Cold* offers a capacious framework for understanding the complexity of temperature's social operation and cultural entanglements: thermal mediation.

FROM METAPHOR TO MATERIALITY

Even though media and cultural studies rarely account for the materiality of temperature, heat and cold are pervasive metaphors in these fields. This is in part because the language of heat and cold evokes density, movement, sensation, and temporality, all of which are fundamental aspects of media ecologies. One of the best-known examples, from which I draw the title of this book, is media theorist Marshall McLuhan's distinction between hot and cool media. Hot media, including photography, radio, and cinema, extend a single sense in high definition. They are filled with data and require little participation from their viewers. Cool media, such as television, cartoons, and the telephone, supply a "meager" amount of information and elicit completion from their users.[27] Inspired by McLuhan but inverting the modes of participation, Phillip Vannini and Jonathan Taggart distinguish between hot energies (such as the heat of a wood stove), which require intense participation and maintenance, and cool energies (such as home heating generated from an electrical grid), which require less bodily involvement.[28] In these pairings, hot and cool are metaphors. Just as hot or cool environments reposition subjects, elicit responses, and affect bodies, so do media and technology reposition, elicit, and affect. Temperature—a mode of environmental description attuned to the rhythm of movement, the composition of substances, and their sensory effects—captures media's capacity to shape time and space. It evokes a sense of media and technology not as stable objects but as dynamic ecologies.

While for McLuhan and others, *hot* and *cold* describe the relationship between a technical system and its users, another lineage of thermal language is more directly influenced by thermodynamics and its "daughter," information theory.[29] Media and culture with high levels of content transmission are often described as conductive: they generate heat and intensity. Concentrated sites of signal exchange are "hot spots." Cold describes the inability to transmit. One photography project, *Screens, Cold*, simply depicts blank, nonfunctioning screens: to be cold is to be off.[30] In these thermal metaphors, mediation itself is always hot; the failure to mediate is cold. The photograph is not hot but, rather, cold and icy as it freezes time and enacts a "spatiotemporal standstill."[31] The cinematograph, with its perpetual movement, is a fiery medium.[32] A foundational author in this vein is Claude Lévi-Strauss, who famously characterizes hot societies as those whose institutions and cultures are designed to accelerate change. Cold societies are those that reduce the effects of historical factors. Heat is aligned with speed, capitalism, repetition, technology, and modernization. Coolness is slowness. The idea that signal exchange is equivalent to heat exchange can be traced back directly to Claude Shannon's theory of information, which advances thermodynamics-inspired concepts such as entropy in the description of communications systems. Today, thermodynamic description often characterizes the extensive transformations of a technological modernity, one that is heating up, increasing in entropy, and ultimately moving toward a heat death.[33]

Along with their frequent use of thermal metaphors, theorists of technology and culture often ground their claims in both the physical transfer of heat and cold as well as experiences of climate. Many argue that the capacity to emit and process heat is a barometer for life itself. "When one gets to the bottom of an animism," Gaston Bachelard claims, "one always finds a calorism. What I recognize to be living—living in the immediate sense—is what I recognize as being hot."[34] Aristotle observes that all living things have a "natural source of heat."[35] And as Thomas Mann puts it, life itself is warmth, "a fever of matter."[36] Whereas, in conceptualizing the human, heat is seen as an essential property of life, in understanding technology, the mastery of heat is what makes humans more than just warm-blooded animals. In *Technics and Civilization*, Lewis Mumford links civilization to the emergence of thermal technologies: wood relieves man of his "servitude" to a cold earth; coal lifts industry beyond seasonal influences; glass hothouses free agriculture from the uncertainty of weather.[37] And for Lévi-Strauss, cooking marks the transition from nature to culture, the process by means of which "the human state can be defined with all its attributes."[38]

Alongside humoral medical theories (which posit temperatures' effects on the body), environmental determinism has permeated Western theories of culture from Hippocrates and Aristotle to the twentieth-century writings of Ellsworth Huntington to contemporary studies of climate change. Environmental determinists posit temperature as a crucial influence in the development of civilization. Montesquieu, for example, writes that people from cold climates are more vigorous, have more courage, and have a sense of superiority, and that people from warm climates commit more crimes, indulge inordinate desires, and are almost "entirely removed from the verge of morality."[39] Exposed to any climate, environmental determinists argue, people adapt. Climatic differences therefore translate directly into social and physical differences. In these writings, heat and cold get linked not only to civilization but also to intelligence, political forms, and race. These discourses are grounded in what Jen Rose Smith describes as "temperate-normativity," the assumption that "proper" civilizations arise in temperate climates and anyone who lingers outside of this zone is aberrant.[40]

Temperature—the measurement of heat—offers a relational system, a means of sorting, separating, and orienting. Whether it is a metaphoric language for media studies or evidence for cultural theory, temperature is harnessed to articulate distinctions: between media, between old and new, between life and nonlife, between humans and nonhumans, between cultures, and between civilized and uncivilized people. In each of these cases, describing things as hot or cold naturalizes a set of social and cultural differences. It marks entities as naturally faster or slower, affective or not, dynamic or static. In turn, these function to naturalize racial, sexual, gendered, and geographic distinctions. The fact that many people believe temperature is an objective measurement system makes these descriptions even more powerful metaphors.

These thermal theories and languages matter. They ground actual practices in the world. As many scholars have pointed out, thermal knowledge actively reproduces racist, colonial, and patriarchal social structures. In a study of French spas and thermal tourism, Eric Jennings highlights how thermal knowledge has been critical in "delineating the non-European 'other' since ancient times" and as a means of legitimizing European dominance.[41] Theories of acclimatization in the eighteenth century, which posited that Westerners could acclimate to the tropics, served as a science of colonization. In the nineteenth century, enslavers used antiacclimation as an antiabolition rationale: white men, they argued, simply couldn't work in the heat. The institutional management of temperature is underscored by these theories and in turn scaffolds hierarchies of race, gender, sexuality, ethnicity, and empire. Labor laws include thermal specifications, and as a result, thermal knowledge permeates work practices. Medical institutions define some bodies as more susceptible to the heat, and as a result, thermal knowledge differentially structures systems of care. Architectural modernity, scaffolded by climate determinism, "inserted a certain type of thermal environment—one that was seen to derive from and to be amenable to inhabitants from Euro-American metropolitan centers—into almost any climatic, social, or political condition."[42] Thermopower materializes in these thermal descriptions—as they shape people's thermal exposures, they become powerful means of racializing, gendering and sexualizing bodies precisely because temperature is perceived as objective, somehow outside of culture and beyond media. In other words, heat and cold are the perfect metaphors to recast social inequities as natural differences. Moreover, as thermal language activates listeners' senses, it layers these naturalized distinctions into a sensory regime; in other words, these distinctions are a way of making racism sensible.

Although a wealth of scholarship exposes the colonial, racist, and patriarchal legacies of environmental determinism, alternative formulations of culture and temperature have been scarce in the social sciences and humanities. An expansive scientific literature explains the physical dynamics of heat and cold, and fields from thermoregulatory science to engineering chart temperature's effects, but these largely avoid culture and embodied forms of difference (even though the science of thermoregulation itself acknowledges the difficulty of predicting thermal response). In this book I argue that the study of temperature cannot be left to the sciences, nor can it ignore culture. Analyses of the relationships between culture, heat, cold, and thermoception must be as complex as theories of images, sounds, and texts, which over the past decades of humanistic and social scientific work have extended far beyond mere determinism.

In one text that has emerged to formulate such a theory, Eva Horn calls for a cultural history of how thermal conditions have been known and imagined. Analyzing climate change discourse, Horn argues that images of inescapable heat offer "the sensory translation of a threat," one that can produce an affective shock and a "phenomenal sensibility" that might translate science—and the large, imperceptible processes of global warming—into political action.[43] Although this description of heat produces a sense of anxiety for some, it also functions to convey "a sense of a newly defined union of mankind as a species" and "a unity of all living organisms."[44] As a result, Horn writes, heat "may serve as a common denominator binding the lives of humans to everything alive that is non-human but equally impacted by rising temperatures and ocean levels."[45] The thermostat maker Ecobee recently put this sense of heat into play in an advertising campaign. The company invited climate change skeptics to a focus group on global warming and then slowly turned up the temperature in the room. Heat itself, Horn and Ecobee imagine, can communicate a forceful message about the danger of climate change.

In contrast, Alexis Pauline Gumbs's speculative documentary *M Archive: After the End of the World* offers a radically different vision of how heat might be felt. Gumbs writes: "while everyone thought of global warming as an external phenomenon, it was happening on the same timeline within. the people on the planet were stars burning out."[46] In this theory of heat, it is not an external force that produces an affective shock but rather something that entangles and reverberates across scales. Later Gumbs continues, "we ourselves made a world too hot for our feet and tried to teach our children to walk in it."[47] The heating of the atmosphere and oceans is akin to the heating of coals, pavement, and ground. Gumbs's writing offers a different politics of thermal communication, oriented not by heat as a universalizing sensorial phenomenon but rather by heat as both an intimate and social process, with felt transformations in sensing bodies that cannot be disentangled from the social world they inhabit.

Media Hot and Cold approaches heat in this latter sense: as a medium of exchange, communication, and violence whose effects will inevitably be entangled on the ground. A close attention to how heat is modulated by thermal media reveals the fundamental fallacy of both environmental determinism and universal thermal communication. Even in the same geographic area, people are subject to vastly different patterns of thermal exposure and are shaped by diverging experiences of their material environments. As I describe in the first chapter, the maids who maintained the fires of aristocratic manors did not experience the cold winter in the same way as the people they served. Their bodies bore the brunt of thermal exposure and carried the responsibility for thermal stability. Who labors—in kitchens or in fields—is determined by the culture and economy of a given society, and the pattern of thermal exposures contours race, gender, and sexuality. In the twentieth century, although Ford Motor Company was the largest employer of Black autoworkers, and even though it appeared that these workers were paid salaries equivalent to those of white workers, they were often given hotter, more dangerous jobs in the company's metal foundry.[48] In turn, as labor economists document, this practice reinforced stereotypes that Black workers were "genetically suited for the hottest, dirtiest jobs."[49] These thermocultures not only extended long-standing histories of allocating thermal exposure based on race but also helped to reproduce race. Temperature is both a gendering and racializing project, especially as it is mobilized within thermal capitalism.

Thermal affects themselves are rarely transmitted directly. They are almost always refracted by thermal media: buildings, clothing, bodies, and technologies. Here is where McLuhan was right—there is no thermal message that is not shaped by the forms that reflect, concentrate, or redirect it. Over a lifetime, thermal media pattern exposures, resulting in senses of temperature that are as variable as ways of seeing. Epigenetic research (albeit conducted using mice and with problems of its own) has suggested that thermal exposures can even alter gene expression.[50]

Some of these thermal mediations are spatial: the dwellings people inhabit, the heating and cooling technologies that regulate ambient temperature, the way that architecture intensifies external temperatures (heat islands, wind tunnels, shade), the places people work (blast furnaces, cold distribution centers). The body itself is a thermal medium: the sense and effects of temperature are shaped by metabolic activity, which itself reflects socially structured access to food and water, as well as cultural practices. The sense of temperature and what heat and cold mean depend on who is cooking, who is standing close to the fire, who is forced to stay still, and who is forced to flee. Thermal mediations are also structured by temporal practices: what times people move, how long they have to labor, how and when they sleep. And yet other mediations are discursive: rooms painted red and blue, thermometer readings, weather reports. There is no unmediated temperature or unmediated climate. Even if heat has a language, there exists a multitude of dialects.

Put simply, climate does not determine behavior any more than television does. Heat and cold are differently accessed, received, sensed, and interpreted, even within a broadcast system. Matter is patterned through ambient transmissions and exposures, all of which are shaped—though not determined—by a multitude of social factors. The exposed subject "is always already penetrated by substances and forces that can never be properly accounted for," Stacy Alaimo observes.[51] And none of these substances and forces can be reduced to a single environmental milieu. While in any given moment, exposure to a draft or forced stillness might produce a sense of cold, across a lifetime or generations, exposures are layered into the body, making up one's sensory perception. Environmental determinist discourse is a violent fiction—one that grabs on to a tiny aspect of an environment at a single moment, erases all of its cultural histories and social inequities, ignores the bodily specificity of exposure, and uses this moment to naturalize a structure of power.

Even though it does not determine thermal meanings, the materiality of thermal media still matters. In the following pages, I describe the primary forms of thermal media, each of which is tied to, though not reducible to, the materiality of their operations.

The first is convective media, such as air-conditioning and ventilation systems. Convective media transform temperature by altering the motion of elements and molecules. While some convective media leverage the physical process of convection (the movement of molecules in a fluid), I also use this term to designate the broader operation of thermal communication through atmospheric forms. The ambient display of heat images can elicit past thermal exposures and operate as a form of atmospheric communication. Convective media are similar to what Yuriko Furuhata and Daniel Barber describe as climatic media, technologies that generate ecological milieus—but convective media often do so specifically through aerial, ambient, or particalized communication.[52] Thermal media, however, are not always climatic. Ice and thermoelectric wristbands, for example, relay a sense of temperature through conduction, the most basic form of heat transfer. Conductive media often capitalize on the physical process of conduction, manipulating temperature through physical contact. These media link bodies into material circuits of heat exchange.

While some thermal media rely on the direct convection or conduction of heat, in which elements and molecules (water, carbon, oxygen, silicon) become transmissions media, radiant media concentrate, refract, and deflect heat waves as a spectral activity. Radiant thermal media also have a distinct physical correlate: all bodies that have a temperature of greater than zero degrees Kelvin (−273.15°C/−459.67°F) emit thermal radiation in the form of waves. These wavelengths generally correspond to the body's temperature. Most entities in the universe emit radiation in the visible light or the infrared part of the spectrum. The sun emits radiation in the infrared range of the spectrum (with wavelengths above 700 nm), as visible light (400 nm to 700 nm), and as ultraviolet light (below 400 nm). Human bodies tend to emit radiation at around 10 μm (micrometers). The plume of a jet's engine emits radiation at around 3 μm to 8 μm (enabling heat-seeking missiles to locate it).

What is distinct about infrared thermal radiation, compared with visible or ultraviolet radiation, is that it's the form of electromagnetic radiation emitted by most objects, bodies, and phenomena that people have contact with in everyday life. It is a form of wave communication in which bodies are immersed and to which they are often responsive. These emissions are not "heat" as we normally know it—they are not a transfer of energy but rather an electromagnetic effect of heat. In a seminal study of radiance, Rahul Mukherjee describes critical aspects of radiant infrastructures: even as they manage imperceptible energy, they are inevitably leaky, and they pull together the public and the private, the spark of development and the threat of contamination.[53] Notably for media theory, radiation moves even in a vacuum—the dissemination of radiant heat thus *requires no medium*.[54] Radiant thermal media often manage rather than simply relay: they organize the multitude of thermal emissions being sent at all times in all directions, most of which are sent in wavelengths undetectable by human vision.

These different material forms—convective, conductive, and radiant—underlie not only thermal communications but communications systems more broadly. Radio transmissions are electromagnetic waves (and as a result, the exploration of early radio dovetails with the exploration of infrared radiation, as described in chapter 4). Telegraph and telephone systems are conductive (and these connect to the development of thermoelectric components that form the foundation for personalized temperature devices, as described in chapter 2). However, while there are many intersections with visual, sonic, and haptic communication, thermal media have their own specificity. Thermal communication, whether messages sent between people in a classic human-to-human approach or messages disseminated environmentally, is defined by several key attributes, many of which diverge from traditional formulations of communications and media theory, especially the information theory descended from thermodynamics.

First, in almost all cases of thermal mediation, communication by heat often appears unintentionally, if it appears at all. People usually do not choose to emit heat in the same way they choose to speak or write words on pages. In many bodies, heat emissions and body temperature are regulated by the autonomic nervous system, which automatically manipulates not only internal heat but also breathing, blood pressure, and digestion. Hot and cold media are less often designed to generate heat themselves than to intensify or modulate apparently "natural" temperatures and already flowing emissions. They tend to work invisibly: the redirection of thermal movement by and large occurs outside the spectrum of visible light. There appears to be no obvious "sender" that is crafting heat signals in acts of volitional agency. As heat and cold are channeled through air, pipes, lenses, metaphors, colors, and buildings, the messages they transmit often seem like ambient properties of a natural environment. Thermal media can produce multiple senders and receivers, all processing and reacting simultaneously. This results in what I describe in detail in chapter 3: a deferral of accountability to the environment, especially by the perpetrators of thermal violence. Thermal media such as a window dispositionally lead us to believe that the sun is heating us rather than feeling warmth as an effect of architects' decisions or our own social practices.

Second, because of their tendency to modulate thermal gradients, communications by heat are rarely discrete, with identifiable beginnings and ends. There are of course exceptions to this rule, several of which I describe in chapter 4: the heat ray telegraph and the directed millimeter rays of the US military's Active Denial System; the infrared transmissions that transport data across fiber-optic cables; and the blast of air conditioners felt upon entering and exiting a shopping mall. But most of the time it is hard to tell where heat ends and cold begins. The perception of thermal stimuli is more often a perception of an environment or an intensity. To engage with heat is typically to engage with an analogue gradient. Thermoception "is not like a thermometer," Vannini and Taggart write. "It is instead an atmospheric attunement."[55] It is not simply that thermal communications have a diffuse spatiality; they often have a temporal indeterminacy.

Third, thermal media often shape thermal subjects through the process of calibration. While calibration typically describes a test to establish the consistency of scientific instruments, it is also used to refer to the standardization of any set of measurement capacities, even those of populations.[56] Thermal media often attempt to standardize bodies as sensory instruments, establishing a normative sense of temperature through normative uses. Just as people learn to see through visual cultures and hear through aural cultures, the sense of heat and cold is conditioned through thermocultures, especially by technologies of temperature control and most deliberately, as I describe in chapter 1, by thermostatic heating systems and air conditioners. This calibration embeds lived routines and material orientations with a sense of hot and cold. As people receive heat signals, they are affected by temperature differently depending on the thermocultures in which they exist. Just as with vision and hearing, the "individual" understanding of thermal communications is actually a situated and collective knowledge.

Fourth, at the same time, felt sensations of hot and cold are always dependent on the particularities of the bodies that are feeling. The reaction of some bodily matter to heat is different from others, however identical they are in composition and however similar their environments. Thermoception is conditioned by past exposures and cultural milieus. Although scientists have begun to document the complexity of these processes, mainstream studies of thermal comfort—the area of research that sets the basis for design of heating, ventilation, and air-conditioning—continue to use models in which culture is largely absent, as are race, gender, sexuality, ethnicity, and history in general. Despite the long histories of bodily calibration described throughout this book and despite the immense amount of research on individual thermal responses, heat effects and perceptions remain indeterminate.

Fifth, this is true in part because the sense of temperature is not merely a perception of an external state. It is a sensation both of external and internal environments, and it thus diverges from other modes of perception that have dominated media studies. "Thermal information is never neutral," Lisa Heschong observes. "It always reflects what is directly happening to the body."[57] Heschong points out that thermal nerve endings are not actually temperature sensors. They are heat-flow sensors. They monitor whether one's body is gaining heat or losing heat. Our sense of temperature, Boon Lay Ong writes, is "a reflection of the energy balance between our bodies and our environment."[58] When people sense temperature, they sense their own energetic state in relation to the world they inhabit. As Marina Peterson points out, since "heat moves spatially and temporally, through animate and inanimate bodies," it connects "the 'matter' of stone, air, and breath, drawing humans and objects together in collective forms."[59] Thermal perception is always a sensation of sameness and difference, while it simultaneously draws the body into such collective formations. Even endotherms, bodies that maintain a "stable" body temperature, are inevitably thermally entangled. Karen Barad argues that entanglement is not "any old kind of connection, interweaving, or enmeshment."[60] It does not presuppose that separate individuals join one another but rather that existence itself "is not an individual affair."[61] What I describe in this book as thermal entanglement is the inevitable intra-action of bodies, beings, and processes with heat and cold. The shape of all matter is fundamentally bound up with thermal conditions.

Last, because of thermal entanglement, temperature fundamentally affects when matter becomes media. Heat and cold alter matter's ability to hold a particular form and its ability to transform. Temperature shapes when matter can facilitate communication—when a window offers a view on a world and when it makes bodies susceptible to attack. Temperature sets conditions for contagion and limits on life spans. No matter what we are composed of, prolonged composure is predicated on assumed temperature. And deviation from this temperature disrupts the rhythms and patterns of existence. Extreme cold incapacitates the paper on which words are printed, the electronic screens where they are displayed, the body and the air in which they are sounded. The overheating of voting machines prompts ballot recounts; the overheating of computers collapses data infrastructures.[62] As a result, beneath all traditional media technologies—textual, aural, and visual—is a set of thermal infrastructures and a thermal regime. The changing climate, as it shapes the conditions of possibility for these thermal regimes, will change not just thermal media but all mediation.

METALLURGY

Over the past several decades, media ecologies have radically transformed. They have become increasingly networked, yet they remain anything but horizontal. New corporations manage global communications, even as these signals transit long-established paths. A massive logistical infrastructure catalyzes media's expansion beyond its former frontiers. Digitization and databases enable new forms of invisible coordination. Protocols are ever more stringent, even as platforms market flexibility and the inclusion of difference. Personalized interfaces channel and redirect affect, generating compulsory habits and insatiable desires. Digital subjects are patterned by centralized systems and stored in massive warehouses. This technical and social reorganization, which has altered the circulation of text, images, and sounds, has also reshaped thermal media. At the same time, the reality and subsequent recognition of climate change has produced new forms of thermal knowledge and new experiences of temperature.

While digital systems and climate change have helped to generate a major shift in thermal mediation, this is not the first time that thermocultures have undergone such radical changes. *Media Hot and Cold* connects twenty-first-century thermal media to the early period of mass media and mass culture, when the audiovisual cultures of cinema, radio, and advertising, as well as the communications systems of telegraphy, telephony, and electricity, had a similarly transformative effect on the sensation of temperature. It is during this period, roughly from the turn of the century through the 1930s, and especially following World War I and the global influenza pandemic, that many of the patterns, technologies, and systems of thermal media in the United States first crystallized. And while in recent decades early twentieth-century thermal media have been reconfigured alongside many other mass media, they continue to set the stage for present thermocultures and those of the future.

The particular cases in this book are not organized chronologically, even though each chapter unravels a genealogy of a thermal medium (including the thermostat, coldsploitation, the sweatbox, the heat ray, the infrared image, and the air conditioner). Rather, I bring these disparate examples together using a metallurgical approach.[63] Academic research often involves retrieving material from the depths of archives, interviews, and texts in order to expose what otherwise might remain hidden. In metallurgy, Gilles Deleuze and Félix Guattari explain, "an energetic materiality overspills the prepared matter, and a qualitative deformation or transformation overspills the form."[64] In contrast to the process of scholarly excavation, this book's metallurgical approach composites histories from already "known" cases, often materials that would not normally be thought together. I subject these examples to a new set of conditions in order to enact a qualitative deformation that activates resonances across and between them. The concepts above, each of which attach "thermal" to an existing area of cultural or media analysis (subjects, attractions, desire, colonization, neoliberalism), are tools of metallurgical analysis: they help to meld areas of study. In this approach, temperature is not an external force to be discovered or documented but something actively entangled in the analytic process. As a result, metallurgy often involves reading sensory contexts in materials that were meant to speak neither about heat and cold nor to each other.

While the overarching project of this book is metallurgical, individual chapters home in on specific genealogies of thermal media. In the first chapter I describe a foundational thermal technology for twentieth- and twenty-first-century thermocultures: the thermostat. This often-overlooked device is a critical interface between people and their thermal environments. It is a site where people learn how to become thermal subjects. They embed what Yuriko Furuhata describes as "thermostatic desire," a desire to posit the atmosphere as an object of human control, often via convective media.[65] This chapter tracks how early thermostats, like early mass media, were developed as part of a broadcast system. Paired with central heating and central air-conditioning, they helped to generate a sense of thermal homogenization and calibrated people as a thermal audience. Thermostatic subjects, I show, were intended to embrace thermal stasis and adopt technologies to reduce thermal fluctuation, and importantly, they were asked to experience the same temperature regardless of race, sex, gender, nationality, and ethnicity. The thermostat is not a neutral interface for a heating infrastructure;

it is a medium for temperature, one that is deeply connected to a cultural need for thermal evenness, stasis, and heteronormative domestic care.

Chapter 1 chronicles how the culture of thermostat use, like other media cultures, changed over the course of the twentieth century. A series of "thermostat wars" broke out over the "correct" temperature. Scientific research challenged the standardized thermostatic subject. In its place, thermal comfort science offered a new imagination of an adaptive thermal subject, one that was shaped by its thermal pasts and capable of adapting to thermal futures. Digital entrepreneurs developed new technologies, such as smart thermostats, that could accommodate diverging thermal desires. This digital approach to temperature encouraged each subject to be in control of their own thermal environment and, implicitly, to be responsible for mitigating climate change. As media transformed from broadcast to microcast, thermoceptive norms shifted from a desire for universal thermal standards to thermal personalization.

While the thermostat often embeds a desire for a lack of thermal sensation, the second chapter focuses on thermal attractions, the sensory pleasures of heat and cold. Thermal media, like all media, have genres. Chapter 2 describes how coldsploitation, a genre of sensational thermal media, exploited the cooling effects of images, sounds, architectures, and technologies for profit. Focusing on the emergence of this genre between the 1910s and the 1930s, the chapter documents a cinematic cycle of enormously popular "snow films." These thermal visions, a form of "haptic visuality" in which "the eyes themselves function as organs of touch," transmitted a sense of temperature.[66] At the same time as these films were released, the lobbies of movie theaters were adorned with elaborate Arctic scenes and radio shows broadcast polar narratives over the airwaves. Coldsploitation media, like so many other exploitation forms, enabled audiences to consume experiences of thermal difference while also legitimating—on a sensory level—projects of colonization. During this period, fans and air conditioners were also deployed in media environments, en masse, for the first time. Collectively these thermal media brought about a new way of consuming coldness. While ice, like newspapers and other nineteenth-century media, was a conductive medium often held in the hand, coldsploitation was a convective medium that shaped atmospheric communication.

Just as we are witnessing the emergence of new thermal subjects, so too are thermal attractions being reshaped in the current moment. The desire for cold temperatures is intensifying. "Last-chance tourism" to the Arctic offers the cold as an increasingly rare and valuable experience, drawing upon a long history of thermal colonization. At the same time, personalized cooling systems that work directly on the body are proliferating. The Embr Wave, a heating and cooling wristband, allows users to give themselves personal temperature adjustments. Smart beds and bedding allow bedmates to each choose their own temperature. Ice vests can be worn under suits, and personal air conditioners are sewn into jackets. These targeted thermal communications are distinct from a century of convective media—using thermoelectric technology and direct touch, they attempt to extract bodies from threatening atmospheres and link them into digitized thermal circuits.

The third chapter turns to another form of bodily calibration—not stasis, adaptation, or attraction, but the enactment of thermal violence. As Kyle Powys Whyte, Heather Davis, Zoe Todd, and Kathryn Yusoff, among many others, have identified, environmental change is a long-standing means of enacting colonial and racial violence.[67] This occurs not only through the dramatic transformation of landscapes but in the everyday modulations of environmental exposures. Taking one example of thermal violence as its focus, this chapter describes the long history and contemporary forms of sweatboxing in the southern United States. Sweatboxing is a violent practice that directly manipulates the body's capacity to mediate heat. In the southern United States in the eighteenth and nineteenth centuries, enslavers weaponized the environment to punish and kill by placing people in wooden boxes in the summer sun or in cold, wet conditions and then restricting food and water. These cultural practices of overheating or freezing persist into the "afterlife of slavery" through policing and prisons.[68] They are advantageous to perpetrators because they offer a way to leverage the indeterminacy of heat effects—the fact that heat differentially affects its subjects—to defer accountability to the environment. It is difficult even to document murders via sweatboxing: many of the people killed have been listed as having died of "natural causes."

Following the long history of sweatboxing into the present moment and the overheating of prisons, this chapter demonstrates how one effect of climate change will be an extension of the human capacity for thermal violence. While some small legal victories have begun to assert thermal rights, such as the right to ice water in extreme heat, changes based on thermal objectivity not only fail to translate into compliance but also miss the ways that heat can be weaponized through a much broader set of practices: forced exertion and labor, the withholding of food or water, and the manipulation of movement and stillness. Instead of adopting thermal technologies or consuming thermal affects, this chapter shows, it is necessary to develop a politics of thermal autonomy: a politics that is attentive to people's ability to regulate and mediate their own position within the thermal world.

The first half of the book focuses on recognizably thermal technologies— thermostats, temperature standards, ice, air conditioners, and sweatboxes— to show how these work as convective and conductive media to distribute heat and cold "content" to thermal receivers: communication here is largely a sense of heat directed from people to people. The second half of the book turns to forms more generally recognized as media, including telecommunications systems, cameras, and computers, to show how deeply entangled traditional media are with temperature. Chapter 4 focuses on the heat ray, a direct beam of thermal communication. In the early twentieth century, alongside the distribution of thermostats and coldsploitation media, many people were experimenting with using heat—rather than light or sound— as part of telegraph, telephone, and television systems. The heat ray telegraph produced signals with superhot searchlights. On the receiving end, a parabolic mirror caught these heat rays, increased the temperature inside a small horn, and ultimately produced a sound in an attached stethoscope. The listener interpreted such sounds, generated by thermal expansions and contractions, as the beginning and ends of dots and dashes. In this thermal medium, heat was generated by a single transmitter and directed to a thermal receiver, and the message was encoded in an existing language: Morse code, English, French, German, or binary code, among others.

This is an example of radiant media that leverages both the spectrum and the processes of convection. Other radiant thermal media of this time include Alexander Graham Bell and Charles Sumner Tainter's photophone, the infrared television, the fever machine, and radio shortwave cookers (that later morph into the microwave). Heat ray media, however, were not taken up as a means of mass communication. It was not until the development of the laser and the fiber-optic cable, a half century after these experiments, that heat rays became the medium of global communication. Today, the most expansive thermal medium is the internet. While the fiber-optic cables that carry internet traffic are largely described as transmitting signals via light, these cables do not use visible light. Data is sent around the world using infrared wavelengths at around 850 nm, 1300 nm, and 1550 nm. Our global telecommunications network is better described as a network of heat rays, a global system of encoding and redirecting infrared waves. Recently, the thermal sensitivities of these cables have come to matter for climate change. New technology has been developed to transform the heat ray transmissions of cable systems into potential sensors for the environment around them: the internet now comprises the largest potential thermometer on earth.

The fifth chapter turns to another thermal medium that has become ever-present with the expansion of digital technology and the COVID-19 pandemic: the infrared camera. While the heat ray telegraph and infrared television failed to take hold in the early twentieth century, infrared photography emerged during this same period as a critical sensing medium, a means of registering bodies and objects from afar. Early infrared images dramatically transformed existing forms of thermal vision and cultivated a new thermoceptive regime, one that recast the world as a landscape of thermal reflectors and that imagined bodies and objects as potential thermal media. Today, digital technologies ranging from satellite imaging to autonomous vehicles rely not only on a web of infrared transmission but also on machinic thermoception. Algorithmic systems aggregate and expand infrared imaging in ever-new ways, offering new means of targeting and visual surveillance. Thermal images are woven through national security and public health. At the same time, infrared cameras are entangled with infrastructures of temperature control, used to define normative temperatures, and tasked with managing a volatile thermal landscape. Through these media, radiant emissions are becoming new sites for extraction, exploitation, and control.

All these digital systems, whether fiber-optic networks or infrared cameras, depend more than any media before them on temperature manipulation. The final chapter of *Media Hot and Cold* is a genealogy of digital media's immense need for cooling. It begins with the first air conditioners, which were made to stabilize media rather than to comfort humans. Without stable temperatures and air-conditioning, phonographs expanded and contracted, radio studios had to retune instruments, and film operators overheated in enclosed booths. Early air-conditioning systems cooled media production and manufacturing, making standardization and speed possible. Air conditioners are also essential to the preservation of media: without stable temperatures, much of media and cultural history would be completely lost. Digital systems build on this long history of media cooling and are today unable to tolerate thermal fluctuation. The trajectory of media's materiality has thus been largely from technologies that could survive vastly different thermal conditions to media that depend entirely on thermostasis.

For environmental determinists, tracking "the coldward course of progress" was a racist naturalization of Western dominance: they insisted that cool temperatures advanced civilization.[69] This coldward course of media is a social project that molds technology in this shape, privileging geographies with stable electrical infrastructure, energy resources, and certain climates. It is a thermal regime that ultimately enacts inequity in media access and institutions. Along with the previous two chapters, this genealogy shows how thermal infrastructures shape communication not only through the sensory forms of heat and cold but by conditioning traditional media operations during manufacturing, distribution, consumption, and preservation. Just as the history of temperature is entangled with the history of communications systems, the history of media technologies is likewise a history of temperature.

By bringing an array of disparate thermal media together—the thermostat, cold air, the sweatbox, the infrared camera, the heat ray, and the internet's infrastructure—this book opens up and broadens the lines of inquiry into thermal transmissions and affects. The content of these chapters will be familiar to some readers. The histories of the air conditioner and the home heating system are integral parts of the history of technology. The sweatbox is a well-known disciplinary technology in the history of the plantation and the prison. Cultural analyses of the Arctic and Antarctic have documented the significance of film in polar exploration. Media history has charted the early developments of the photophone and the television. Digital media studies has exposed the coldness of the data center. Histories and contemporary analyses of architecture have explored the critical importance of temperature in environmental design. As part of this book's metallurgical approach, I draw together foundational accounts from all of these areas, along with trade and industry literature, ethnographic studies, and scientific research on temperature.

Ultimately, the qualitative deformation I hope to achieve by bringing these disparate cases together is that of your own skin and the skin of others. I hope that you will feel these words. I hope your attention will be redirected to the multiple and contradictory ways that temperature affects bodies. I use the words *heat* and *cold* loosely to evoke sensation and affect. I write to a "you" without knowing who you are but knowing you are somehow sensitive to temperature. If understanding media entails understanding the ways we are positioned, hailed, and attuned through the material configuration of communicative forms, then the true project of this book is to leave you thermosensing differently.

IN 2020, THE COVID-19 PANDEMIC consolidated trends already emerging in thermal media. Infrared imaging cameras long used in medical and security operations were deployed as part of health surveillance systems in warehouses, transit hubs, and schools. Infrared spectroscopy was adopted for diagnosing COVID-19. Thermal images revealed the hidden trajectories of microdroplets through the air.[70] Social and economic activity were quickly transduced into infrared signals and zoomed through the internet's chilled infrastructures. Conductive thermal media, which hooked bodies directly into circuits of exchange, became ever more appealing in a moment of atmospheric contagion. At the same time, wildfires raged, ice sheets melted past the point of no return, temperature records were broken, and popular media offered stories on the "profoundly unequal" impacts of extreme heat.[71] More than ever, thermal media appeared as a solution to social problems—they would be indicators of health, a means of sustaining economic activity, and a way to manage bodies.

The pandemic also sparked a new wave of thermal research on the connections between weather patterns and viral spread. Some studies drew correlations between daily maximum temperatures and daily incidence rates of COVID-19, pointing out that diagnosed cases increased as the outside temperature decreased.[72] Others suggested that higher latitudes and colder climates fostered conditions for viral transmission.[73] Drawing from prior work, much thermal research operated within long-standing frameworks of environmental determinism. A century earlier, following the 1918 influenza pandemic, Ellsworth Huntington argued that of all possible factors, from a city's geography to its inhabitants' physiological condition, "the only one which shows any conclusive causal relation to the destructiveness of this particular epidemic is the weather."[74] Like the results of earlier studies, many of the climatic connections of COVID-19 relied on daily temperature readings or climatic averages, with little attention to the cultural entanglements of temperature that shaped exposures on the ground: architectures' differential mediation of heat; the ways that hot weather fostered gathering in cooling centers; the social stratifications of thermal reception.

Media Hot and Cold composites a sensory impression of this thermal regime, which was already in formation in the late twentieth and early twenty-first centuries but which became strikingly visible as temperature shaped pandemic conditions. The future of temperature, this book shows, will be deeply enmeshed with digital media and articulated as a reaction to climate change. As the founder of one AI-based thermostat states, "We believe that together, we can change climate change."[75] Digital devices enable some individuals to micromanage their thermal environment, mitigating atmospheric effects while at the same time naturalizing a sense of neoliberal thermal control. Thermal technologies are legitimated by new strains of thermoregulatory science that suggest humans can take responsibility for their thermal well-being. At the same time, climate change expands the capacity to profit and enact violence using thermal affects. Infrared communications are dramatically altering the cultures of agriculture, security, and public health. Even as these thermocultures are described as "green," they capitalize on thermal instability and are enmeshed in an industrial-technological shift to distributed, high-energy computing and manufacturing that itself accelerates climate change.

Climate change and large-scale technological systems such as the internet are often discussed as global phenomena. They appear to exist beyond the scale of everyday life, human perception, and local politics. "Since we experience weather, not climate," Wendy Hui Kyong Chun writes, media are essential to registering global climate change and "opening climate up to cultural inquiry and political mobilization."[76] As many scholars recognize, climatic media support weather forecasting and progressive policy changes at the same time as they scaffold massive geoengineering projects. However large-scale coordinated changes alter temperatures on earth, they will do nothing to ameliorate long-standing practices of thermal violence. In other words, even if climate change were miraculously reversed, thermal inequities and harm would persist as they have for centuries.

In contrast, this book, attuning to the ways that people and media are entangled with heat, directs attention to the much more minute and intimate effects of thermal shifts. Climate changes will alter everyday rhythms and movements, changing the capacity to transmit or hold still. The power of temperature as a means of sorting, categorizing, and distinguishing will increase. Its ongoing entanglement with processes of racialization will continue. In ever-new thermal environments, there will also be new potentials and new blockages for mediation. The thermal resources needed to host internet traffic will become more scarce and the geopolitics of telecommunication will shift. New sets of thermopolitics are already emerging, not only around the Arctic ice and rising seas, but also in conflicts over the ability to create and maintain thermal zones. As Sheila Watt-Cloutier evocatively explains in *The Right to Be Cold*, Inuit culture and economic independence "[depend] on the cold, the ice, and frozen ground," and "the great shifts in temperature and weather patterns [are] upending an entire way of life."[77] Climate changes will reverberate across thermocultural forms.

Media Hot and Cold offers a framework to address such shifts, analyzing the thermal media with which people are already engaged. I show in the following pages that social politics are *already* climate politics. Data protections would help to ensure thermal autonomy. Prison abolition would mitigate thermal violence. Access to local thermal media enables more community control over local thermal capacities, whatever the global thermometer reading. This ranges from the ability to alter thermostat settings to the insulating potential of buildings and fabrics. It extends from the recognition of vernacular thermal perceptions to the creation of new and equitable thermal practices. New thermal media, and even writing itself, can be used to calibrate different kinds of thermal subjects. Critical temperature studies need not only analyze from a distance. It can work generatively to reorient affective responses to the world of heat and cold.

PART I

1

THERMOSTAT THE THERMAL SUBJECTS OF
BROADCAST TEMPERATURE

Along with the sound of keyboards and the off-white glow of fluorescent lights, the bodies of office workers register heat and cold. They feel cool air funneling out of a vent. They are subject to radiant heat amplified by the building's windows. These ambient thermal communications are engineered to raise productivity. Even in the 1910s, air conditioners were marketed with the promise of increasing worker efficiency. A century later, 91 percent of the floor space in United States commercial buildings would be intentionally cooled, much of it to the 20°C–24°C/68°F–76°F range suggested by the Occupational Health and Safety Administration. Despite decades of research and experimentation, the ideal temperature for office space remains under debate today. A Cornell ergonomics expert found that typing errors decrease when the temperature is raised to 25°C/77°F.[1] Mark Zuckerberg set the Facebook office thermostat at 16°C/60°F.[2] The determination of optimal temperatures—and the ability of thermal capitalism to profit from thermoception—is an ongoing struggle, and it relies on the definition of an ideal thermal subject.

For much of the twentieth century, thermal capitalism assumed that its subjects were thermostatic—desiring stable and consistent temperatures. And yet time and time again, the "ideal" temperature wasn't received as intended. Workers read the air moving past their skin differently. For some, the setpoint of the thermostat was too high; for others, too low. One person turned the temperature down; another turned it back up. These thermostat wars took place throughout the late twentieth century, but they exploded in public discourse in the summer of 2015—to that date, the hottest summer on record.

The explosion was sparked by a publication by two Dutch scientists that revealed how standards for office air-conditioning, deployed around the world as thermal norms, were based on a thermoregulatory model of a forty-year-old man who weighs 154 pounds. The article concluded that offices should "reduce gender-discriminating bias in thermal comfort."[3] The research was publicized in the *New York Times* and was ultimately reduced to a tagline: air-conditioning is sexist.[4] The *New Yorker* suggested that more "thermally egalitarian offices" would bring about changes in work culture.[5] A campaign was launched "to close the gender climate gap" in Canada.[6] In conservative social media, memes featured air conditioners that announced, "I hate women!" The thermostat wars would eventually circle back to productivity. Four years later, yet another study found that women solved math problems better at warmer temperatures and men solved them better in the cold.[7]

But in 2015, the study and its media coverage catalyzed a second thermostat war, a symbolic one that was waged over the meaning of heat and cold. On one side, advocates of thermal difference criticized the normative temperature standards anchored by white male bodies. They argued that people thermosense differently and, as a result, diverging experiences of temperature needed to be accommodated. For others, thermal reception remained outside culture and beyond media. Calls for thermal accommodation became easy targets for parodies of liberal inclusivity. At its core, the symbolic thermostat war was a struggle over thermal subjects. It was a struggle over whose bodies anchored thermal standards and defined normative thermal practices. Unsurprisingly, digital media stepped in with a solution. Several start-up companies developed technologies that seemed to bypass the need to reconcile thermal difference. Their thermal apps got rid of broadcast temperature—the practice of thermal distribution in which an ideal set of thermal communications (say, 20°C/68°F) was sent from a central source to a mass of receivers. Without a broadcast system, there would be no need for a thermal standard.

This was a critical moment in which thermopower was questioned and contested. Through much of the twentieth century, the dominant mode of thermopower, like so many other media forms, calibrated subjects through mass technological systems, including central heating and central air-conditioning. Alongside other forms of mass media developed in the early twentieth century, broadcast temperature systems attempted to uniformly heat and cool entire buildings, floors, and environments. Standard temperatures were defined for these spaces. Architectural historian Luis Fernández-Galiano argues that these infrastructures produced a culture of thermal homogenization, one that paralleled the visual and technical homogenizations of modernity.[8] The normative thermal interior, Daniel Barber shows, not only was a product of mechanical systems but was intertwined with architectural modernity itself.[9] Like other forms of mass communications, the system of broadcast temperature was structured as one to many. All people would encounter the same environment and, in turn, would be shielded from encounters with the same unwanted temperature. The emergence of civilization is often described as the ability to control temperature. The emergence of modernity is correlated with thermal uniformity and regulation.

Although researchers have investigated heating and cooling systems as architectures of thermal uniformity, few have described the subjects calibrated as part of these projects.[10] The broadcast temperature system cultivated a thermostatic subject: one that was supposed to desire stasis and continuity, avoid fluctuation, and operate as part of an automatic control system.[11] People were addressed as members of a mass and evacuated of any cultural markers, their bodies imagined as blank slates immune to the effects of time or geography. Even though few people met the ideal of these technical practices and standards, the thermostatic subject marked a departure from an environmental determinist imagination that presupposed a population already affected by its thermal environment (although environmental determinists played a key role, as Marsha Ackermann recounts, in helping to spread air-conditioning systems and comfort standards).[12] The thermostatic subject, in contrast, was imagined as an ideal and standardized receiver, a raceless, genderless, and sexless body that felt temperature in predictable and manageable ways.

This chapter describes the emergence of the thermostatic subject both in technological practices, standards, and regulations and through an array of messages about temperature that attempted to neutralize thermal differences and erase thermal pasts. The first section of this chapter, drawing from the advertisements and archive of the Minneapolis-Honeywell Regulator Company, describes the social origins of the thermostat and its expansion into the twentieth-century American home. One of the key operations of the thermostat—the user interface for a building's heating and cooling infrastructure—is to divorce temperature from energy. Before the deployment of the thermostat, the heat and cold of the interior environment depended, by and large, on physical engagements with fireplaces, furnaces, and other technologies of thermal production. Because it detaches inhabitants from these infrastructures and the labor necessary to maintain them, the thermostat inscribes temperature as a manipulatable representation. For those who set it, the thermostat becomes a means of controlling a broadcast thermal communications system, albeit one confined to their own home. In the process, heat and cold are recast as affective states to be domestically consumed and controlled from an interface. This chapter brings together advertisements and industrial discourses that convey normative thermostatic practices in order to show how, early on, the device attuned its subjects to thermal regimes of evenness and continuity, situated them in a historical set of class relations, and positioned thermal communication as an extension of heteronormative domestic care. From its beginning, this chapter reveals, the thermostat has helped to produce temperature as a gendered phenomenon.

The thermostatic subject was calibrated not only in the operation of heating and cooling interfaces but in also the movement through thermostat-controlled public and commercial spaces. In the 1910s, the installation of ventilation systems in institutional and industrial buildings in the United States prompted the development of thermal standards—a set of ideal thermal messages. This chapter's second section traces the origins of these standards in scientific and engineering communities. In the comfort zone chart of the 1920s, heating, ventilating, and air-conditioning (HVAC) researchers developed an image of the universal thermal subject that would respond in predictable ways to the sense of cool and hot. The "ideal" temperature, reframed as "comfort," naturalized a set of dominant thermal desires defined by a degree range and a *lack* of thermal sensation. The comfort zone chart and its standard subject were subsequently encoded in architecture, building operations, and a multitude of mass thermal-control technologies.

The early history of these practices—the cultural programming of thermostat users and the definition of a thermal comfort standard—reveals attempts to standardize thermal perception and flatten thermal differences during the first half of the twentieth century while implicitly reproducing long-standing structures of gendered thermal perception. Despite such efforts to provide an official language of thermal communications, thermoception remains vernacular. The receiving body is still a body, patterned through the specificity of repeated exposures, which are themselves shaped by social practices and cultural history. Overheated and chilled people in homes, theaters, and offices have questioned normative thermal regimes from the outset. Just as the thermostatic subject was formalized in thermal comfort research, discourses of the "thermostat war" began to consolidate these rogue affects. While early thermostats were marketed as easing gendered tensions, from the 1950s onward, thermal difference was often articulated as a gendered conflict. At the same time, more complex thermoregulatory models and computational systems destabilized existing thermal assumptions and introduced subjective opinions into thermal comfort equations. As they proliferated variables, researchers began to question the mass nature of thermal deployment altogether.

The alternative that began to circulate was a personalized model of thermal comfort. In the early 2010s, products such as the Nest thermostat and the Comfy app began to deploy this new approach to temperature control. The final part of this chapter tracks the rise of thermal personalization and the corresponding emergence of the adaptive thermal subject, who is imagined, in a neoliberal mode, to be able to individually adapt to existing thermal conditions using new technologies. This is in turn cast as a feminist endeavor and a "lean-in" to corporate power. If the project of modern architecture, business, and institutions was to create thermal uniformity based on a universal standard of comfort, these new thermal technologies mirror the personalized, user-centered orientation of the rest of the audio, visual, and textual mediascape. The thermostat war of 2015 ignited these latent tensions and catalyzed investment in technologies that reject the thermostatic user. No longer, developers claim, is there a singular subject who needs to receive a single message. And yet even as these technologies promise autonomy, they ask users to both internalize their own thermal problems and to cede control to a broader thermal network controlled by utility companies.

Sketching out this transition, this chapter reveals that this shift in thermopower is fundamentally a shift in a form of mediation. Early twentieth-century thermocultures were shaped by broadcast technologies, anchored by a standard, universal subject, and oriented by a homogenized thermal landscape overseen by a thermal caretaker. Early twenty-first-century thermocultures are built on microcast and personalized organizational models. They imagine a neoliberal and flexible subject, one who embraces variation along with personal thermal responsibility. Proponents of thermal personalization claim that new technologies will help to alleviate climate change, empower women, and mitigate class tensions. These claims suggest that thermopower exists in digital self-control.

What this chapter reveals is that thermopower is not simply the ability to directly alter temperature—say, by turning up the heat. It is the capacity to shape embodied movement and subjectivity through the manipulation of thermoceptive regimes. Thermopower works through the calibration of bodies' sensory expectations and the establishment of normative relationships to temperature. This entails not only subjecting people to repeated thermal experiences but also layering these experiences into a social landscape and naturalizing them through ideologies of heat and cold. Offering access to thermostats doesn't grant people thermopower: they are still subjects whose movements and agency are systematically constrained. By focusing not simply on the energetics of heat but on the calibration of bodies, these cases show how technologies now code thermal difference as an underlying personal preference. They create an affective milieu in which an on-demand climate change can be controlled and consumed. In users' skin, anthropogenic climate change is normalized. The adaptive thermal subject is thus called into a position of responsibility for their own bodily temperature, abdicating any responsibility for the thermal care of others. In the process, the larger structure of thermopower is obscured.

SEPARATING TEMPERATURE FROM ENERGY

On a British estate, the scullery maid woke up early in the morning. In the summer her attic room would be hot, and in the winter it would be cold. She would descend to the kitchen to light the morning fires. Throughout the night, a girl had been traveling between the bedrooms of the house, tending their flames. For the aristocracy's sons, these girls carried the promise of warmth. Another maid maintained the drawing room fireplace. These women's bodies were sensors. They were calibrated to act when it became too cold, to mobilize others to bring wood or coal, to start and maintain thermal technologies. They organized the labor of heating and cooling so that the aristocratic family would enjoy relative thermal consistency. And yet, even as they operated as part of a thermal infrastructure, their bodies were far more susceptible to heat and cold than the bodies of those they served. Their rooms were located in attics. The gardeners slept in huts close to the greenhouses they were responsible for keeping warm.

While the maids, the gardeners, and the aristocracy all lived in the same climatic zone, their bodies were differently calibrated in relation to their social position.

Thermostats are devices for environmental regulation that combine a sensing capacity with a heating or cooling apparatus in order to maintain a specific temperature—the setpoint. This instrument was first developed to regulate industrial processes in the factory, to "mediate between the origins of power and the output of products."[13] But long before they became an industrial technology, thermostats were women. They were maids and wives and mothers whose gendered labor of domestic care involved anticipating the thermal needs of others. In a history of home heating, Sean Patrick Adams observes, "Day-to-day care of the hearth was considered a woman's work, as it always had been."[14] With the introduction of coal in the nineteenth century, the responsibility of heating became "more dirty, demanding, and time-consuming," and even if allocated to servants, "the task of cleaning and polishing stoves almost always fell to female staff."[15] Regardless of technology, temperature has a class and a gender division. Bodies are differentially exposed to debilitating thermal conditions. Some people are sustained and made comfortable through the thermal labor of others.

When the first thermostat technologies were deployed for human comfort, they automated existing thermal labor practices, formalizing a division of temperature from energy that had long permeated class- and gender-based thermocultures. Warren S. Johnson, professor at Whitewater Normal School in Wisconsin, frustrated by his inability to control the temperature of his classroom, patented an "electric tele-thermoscope" in 1883. This early thermostat was simply a communications system that rang a bell when a specified temperature was reached. This notified a "fireman" in the basement to open or close the furnace damper. Although the thermostat is often described as a device that merely regulates temperature, the key operation of this regulatory system was social. Prior to the tele-thermoscope, teachers would have to seek out a janitor, or a janitor would have to enter the classroom "to determine if it was too hot or too cold and then adjust the dampers in the basement."[16] The janitors' skin, like the skin of the housemaids, was deployed as a thermal sensor to maintain a class-based thermal order. Like the housemaids' work in the house, these janitors were exposed to the heat of the furnace and the cold of the winter as they labored to keep the white male students in the classroom warm. Johnson's innovation was less of a means of keeping control of temperature than it was a means of segregation, replacing the janitor's active sensing with a thermometer.

Twelve years later, Johnson patented another technology, one that automated the other parts of the system through a different form of network communication: pneumatic tubes. Johnson's "heat-regulating apparatus" used the tubes to send multiple "air signals" from rooms across a building. The signals automatically controlled a valve on a steam heater, a water heater, or a furnace damper. This eliminated janitors from the communications system altogether, even though they still delivered wood or coal to the furnace. This substitution was pioneered years prior in Alfred Butz's 1885 "damper flapper," which connected a thermostat to a pulley system in order to automatically open and close the furnace damper. These two technologies, and the companies established to sell them, formed the foundation for the modern thermal-control industry. The Johnson Electric Service Company, now multinational conglomerate Johnson Controls, shaped heating regulation over the twentieth century, and Butz's company, which merged with Honeywell Heating Specialty Company in 1927, remains the leading thermostat manufacturer to the present day.

Although these two thermostat systems are regularly narrated as the origins of thermal control, they automated only a small part of an existing human-thermostatic system. Instead of maids or janitors using their skin to establish an even temperature for others, the thermostat removed the sensor from a human body. "Replacing servants with services was a persistent theme in the literature on thermostatic systems," Michael Osman observes in a study of the device.[17] These systems enabled the beneficiaries of thermal modulation to further distance themselves from the labor and laborers of thermal maintenance. As Neda Atanasoski and Kalindi Vora have shown, discourses that highlighted the "replaceability" of particular kinds of human labor were both racialized and gendered—the "tasks deemed automatable . . . were regarded as unskilled and noncreative—work that could be done by the poor, the uneducated, the colonized, and women."[18] Maintaining a thermal environment was one such "unskilled and noncreative" form of labor.

Both of these technologies were developed in the northern United States, where even in the early twentieth century many households were still heated by coal-burning stoves. Central heating, however, was increasing. In hot-air furnaces and steam-heating installations, heat wasn't directly transmitted to bodies from a stove; instead, it was distributed through a building using air or water. These convective media, even if they remained a comfort largely enjoyed by the wealthy prior to World War I, became a prime site for thermostatic regulation.[19] And even as gas and oil heating systems were being developed, coal in this period remained "the domestic fuel of choice."[20] Regardless of heating system or fuel choice, for any household without a thermostat, temperature change was an embodied and manual process typically undertaken by women. It required a visit to the stove or, in the case of a hot-air furnace, the basement, and the regular stoking of coal.[21] The thermostat promised to automate part of this labor.

As thermostats were deployed as part of centralized heating systems, in which warmth was broadcast from a single heating plant to multiple parts of a home, they worked as intermediaries to the furnace. Temperature became a degree manipulatable through a technical interface. To know temperature was to know a number. To change temperature was to change a representation. To sense temperature was to feel an interface. Hooked into a system of thermostats, circuitry, and ventilators, Osman writes, people oriented toward the domestic sphere as a domain of management, "an internalized network of control."[22] No longer connected in the same way to energy or labor (whether one's own labor or the labor of servants), temperature as representation opened up a multitude of possibilities for what thermal sensation might mean. And in the process, it introduced a new user orientation: the thermostatic subject.

The first decades of thermostat advertisements helped users make sense of these new technologies and normalized particular practices of thermal control. Thermostatic subjects are defined by the need for evenness—in direct contrast to perceived variability and disruption. Thermal evenness was not always an aspiration of heating systems, nor was it an invention of thermostat manufacturers. For many years, fireplaces had provided uneven heat, often baking one side of a room while freezing the other. "A person warmed by radiant heat is, however, always unequally warmed," states Thomas Tredgold's 1824 text on heating and ventilation, and "it is certainly an overstrained idea of comfort to suppose an absolute equality of heat desirable."[23] As stoves came to replace fireplaces in the United States, especially in the 1820s and 1830s, and as coal came to replace wood, the perception of ideal heat shifted. The relatively new concept of comfort and the corresponding notion of thermal evenness were critical to the transition to coal itself in the early nineteenth century.[24] Later on, evenness would become a selling point of oil systems.[25]

While evenness was generally tied to the marketing of heating systems and fossil fuels, it was particularly significant in the culture of thermostat operation. Even temperatures were sold not simply as an ideal thermal state but as a means of easing gender difference. With the thermostat, inhabitants would no longer be subject to either variable temperature or heterosexual conflict: they would be offered stasis by off-loading thermal care onto a centralized system. A series of advertisements from 1896 and 1897 for Johnson's Furnace Draft Regulator features many of the tropes that would anchor thermostat sales into the twenty-first century: the thermostat "regulates itself" and works automatically "without taking up your time or attention" (it is autonomous), "saves its own cost" (it is economic), "saves health," and increases the "family wealth" (it maintains the family).[26] In one image, a father, mother, and two children sit around not a fire but the thermostat, which occupies a central place in the household and is depicted as a light in the darkness (figure 1.1). The text of the ad, with the headline "The Husband Explains," features a wife asking her husband to account for his "even temper." Her husband explains that since the regulator saves him from annoyances and gives him an even temperature, he now has an even temper and wishes to remain at home. The regulator, he says, "completes our domestic happiness."[27] Another advertisement from the same year promises that this regulator would get the "extremes under control."[28] Users were encouraged to see temperature control as a means of bonding together as a family unit, shielded from any internal differences via insulation from external variability.

As these instruments spread, thermostat operation became a new form of home maintenance for women. In turn, women became key targets and subjects of thermostat advertisements. One 1925 ad describes a husband's frustration with variable temperature—and with his wife's mismanagement of comfort. "Almost every time he came home the house was either hot as a bake-oven or cold as an ice-box, seldom comfortable," it claims.[29] The wife wasn't incapable, the Minneapolis Heat Regulator Company argues; it was simply impossible to keep the temperature uniform. Comfort just couldn't be supplied by a human. A few years later, the Minneapolis-Honeywell Regulator Company launched a series of ads that suggested thermostats could be made be "keenly interesting to women," the "'powers that be' influencing every sale."[30] In one image, the text beneath a man reclining in his chair reads: "Once in a while you find a man who likes to get up early and go down in the cold to start the fire." This man wouldn't mind doing "janitor work" or having his wife do so to maintain the heating plant. Most people, the advertisement claims, would instead choose a thermostat to give them "a lifetime of automatic service."[31] In another ad, an image of a woman looking at her husband is captioned: "Just think of all the drudgery our heat regulator saves me." The text goes on to claim that since the "fire tending responsibil-

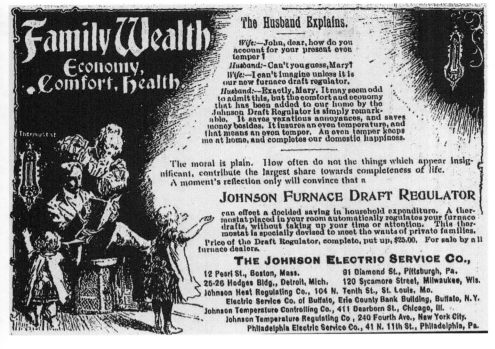

FIGURE 1.1. An 1896 advertisement for the Johnson Furnace Draft Regulator claims that temperature regulation will keep husbands at home. Honeywell, "Family Wealth," *McClure's Magazine*, September 1896, 162.

ity" has been lifted from the wife's shoulders, she is now "free to go visiting or shopping or to a matinee whenever she wishes, knowing that the house will be warm and comfortable."[32] Thermostatic subjects, and the evenness they are sold, were bound up with a transition to a consumer society and the emergence of "domestic engineering."[33]

Of course, even with heat regulators, the labor of heating never really disappeared. Controlling the temperature still required someone to deliver fuel, set the temperature, and maintain the system. What was sold was an idea of comfort as freedom: from having to stoke the heat and from housework itself. But as was true for many of the technologies Ruth Schwartz Cowen identifies in the classic text *More Work for Mother*, the thermostat ultimately intensified the gendered responsibility to take care of the family's thermal needs. At the same time, as Osman observes, heating technologies helped redefine the domestic sphere as "complementary in its technicality to that of industrial production."[34]

Heat regulator companies came up with more and more thermal needs to satisfy. They developed new imaginations of comfort, each of which could be fulfilled by the purchase of the latest automatic device. Thermostat companies also helped imagine new thermal threats, using "fear" of incorrect temperatures "as a means of attack."[35] In the late 1920s, Minneapolis-Honeywell relayed in its *Regulator News*, a publication for its salesman and distributors, that the home "is always too hot because the housewife is afraid it will get too cold."[36] The company embarked on a campaign against overheating, with a cautionary tale featuring "Little Jimmy," a "pathetic . . . cold-choked little chap" who has been debilitated by overheated houses. Minneapolis-Honeywell suggests, "The child-mother appeal always plays on the heartstrings of parents."[37] But the company cautions that sellers must be "tactful," since "such arguments are dynamite if overdone."[38] Launched with a corresponding booklet, "The High Cost of Overheating," the campaign transmitted not only new thermal fears but a wealth of scientific knowledge to support them. At the same time, women were still described as thermally vulnerable and in need of heating protection. One man reported his troubles to an Illinois "building comfort service": "I just simply don't seem to be able to get my house warm enough to satisfy the Mrs."[39]

The programmable thermostats of the 1930s, which allowed the automation of variable temperatures (the user could set them to turn up the heat during the day and lower it at night), were still described in terms of the evenness and continuity they produced. With its sensing capacity, the thermostat "'feels' for coming changes in temperature" and is able to offer "*leveled heat* for the first time," avoiding "dangerous dips" and "wasteful bumps."[40] One Honeywell ad featured a glass tipped to the side with its water remaining level—in the background cascading temperatures fall from hot to cold, in a graphic that echoes the stock market crash (figure 1.2.). A series of Honeywell advertisements from 1934 foreground the thermostat's agency, automation, and predictive capacity: "Almost human," they read; amalgamating the maid, janitor, and wife, the Chronotherm can "put your heating plant to bed when you retire."[41] Another ad promises that the Chronotherm will "forecast your weather inside a year ahead." By 1937, Honeywell was describing its Acratherm as having "the power actually to sense indoor temperature changes before they occur," bringing "the blessing of stabilized heat."[42] In a decade when hot, dry summers in the United States intensified the Dust Bowl, the great heat wave of 1936 ("the most severe protracted weather event in North America's recorded history") resulted in more than five thousand deaths, and a cold wave dropped temperatures

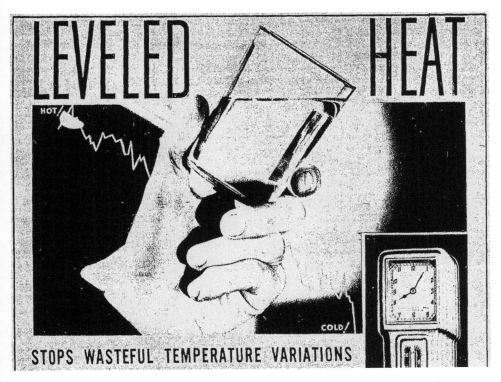

FIGURE 1.2. An advertisement for the Chronotherm offers "leveled heat" as a solution to volatile temperatures and economies. Honeywell, "Leveled Heat," *Detroit News*, September 2, 1934.

across the country, thermostats were marketed not only as the means of withstanding variation but as predicting and acting in advance to ensure thermal continuity and stabilization.[43]

Over the course of the first half of the twentieth century, each new thermostat was guaranteed to bring more automation than the last (an ad run in 1951 and 1952 for the electric clock thermostat promised to finally make heating "*completely* automatic"), offering its users a fantasy in which they could finally and completely disconnect their thermal environment from their own body's energy expenditure.[44] And while the meaning of home and family shifted across the late nineteenth and early twentieth centuries— from the moral concerns of the 1890s to the Depression-era concerns with stability and through the anxiety around the nuclear family of the 1950s— across these periods, the thermostat continued to mediate ongoing gendered

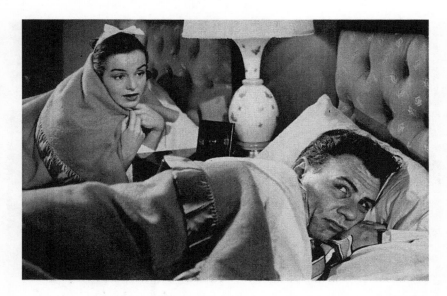

"No! It's your morning to turn up the thermostat!"

FIGURE 1.3. An advertisement for a Honeywell electric clock thermostat promises that husbands will be able to stay in bed longer in the morning. Honeywell, "No! It's *Your* Morning to Turn Up the Thermostat!," *Life*, November 5, 1951, 63.

domestic conflict. Take, for example, the similarities between the ad for Johnson's Furnace Draft Regulator (figure 1.1) and an electric clock thermostat ad of the early 1950s (figure 1.3): the latter depicts a wife and husband, each curled up under separate covers, ostensibly fighting over who will turn up the heat. The text below the image reads, "Soon the husband will get out of his nice, warm bed—and shiver down the hall on icy floors." Honeywell promises that its "sensitive" appliance will take over this "thankless, wintry-morning chore," keeping the room temperature uniform, "no matter how the weather changes." As had been true since the late nineteenth century, the thermostat was intended to bring together the family and ensure heterosexual stability through thermal stability.

The early thermostat offered its users the ability to mediate indoor climates with interfaces, buttons, and dials. Yet across the first half of the twentieth century, this increased capacity for thermal change was sidelined in favor of a focus on achieving stasis and reducing fluctuation. In many ways, this is representative of a broader cultural subjectivity established through centralized systems and mass media. But as advertisements, booklets, and signs on store counters transposed this logic to thermostat-based communication, they restructured common understandings of the manipulation of heat and cold, informing users of "proper" thermocultural practices. Evenness had to be sold. The primary site of thermal difference was defined as the boundary between the inside and the outside of the home. The primary thermal struggle was to maintain this boundary. Inside the home, inhabitants were not struggling over *whose* thermal comfort would be accommodated: the family was a thermal unit (even if it was almost always the husband's thermal comfort that was a concern). As a result, in this period, the thermostat became part of the perceived apparatus of care and a means of maintaining heterosexual relations and reproductive futurity. It was a tool of both family management and wife management. It was humanized as a kind of personal maid and invisibly embedded a long history of racialized and gendered labor relations. Outsourcing this labor—and, indeed, separating temperature from energy—was the precondition for even temperature as a preference, a representation, and a consumable affect, none of which was a natural or universal thermal desire.

THERMOSTAT WARS

The social, architectural, and technical changes of the 1950s, even as they scaffolded the discourses of thermal evenness and thermostatic care, also set the stage for the fracturing of the thermostatic subject. By the mid-twentieth century, the Minneapolis-Honeywell Regulator Company had achieved a degree of saturation in the thermostat market. Tasking the advertising agency Foote, Cone, and Belding with a market study, Minneapolis-Honeywell learned in 1951 that three out of four consumers had no complaints about the heating conditions in their homes. Their environment was already comfortable. As a direction forward, the advertising agency suggested, "We must create consumer dissatisfaction with present heating conditions."[45] The agency proposed a "dissatisfaction campaign." The key to this, it argued, was to shed light on small scenes of discomfort—and the task was to make people "realize this discomfort" that would otherwise remain unconscious.[46] One ad in the campaign, in striking contrast to prior images of families bonded through thermostatic comfort, positioned the thermostat itself as an interloping mistress. "This house isn't big enough to hold the three of us!" the wife exclaims to the husband.[47]

One solution to the problem of market saturation was to turn the thermostat into an aesthetic object. While earlier thermostats were marketed as "an attractive electric clock," in 1953 Honeywell released the T-86 thermostat—described as a fundamentally new design and improvement on previous "ugly" thermostats.[48] *BusinessWeek* proclaimed that it represented the "first major change in basic styling of thermostats in 70 years."[49] Honeywell ads displayed T-86s in an assortment of different colors as objects of aesthetic appreciation (figure 1.4). The T-86 offered a new aesthetic of temperature control in which roundness and color indicate a capacity for temporal and visual synchronization of the thermal environment.

While the success of the T-86 would shape thermostat design for years to come, the advertising agency meanwhile proposed something much more dramatic than just aesthetics: it would create a "new concept of comfort," zone control.[50] By the 1940s zone heating had long been used for commercial and institutional buildings where multiple rooms or tenants required separate thermal systems, and this concept began to gain traction as a model for domestic heating design. (Prior to this point, "zone heating" was an obstacle to be eliminated. One advertisement told potential buyers that their regulator would overcome "3-zone heating" and offer even temperature in its place.[51]) In 1943, framing the development of domestic zone heating as part of the wartime effort, Harold W. Sweatt, the president of Minneapolis-Honeywell, announced, "Controlling heat furnished to different parts of the house so as to obtain healthful living conditions with lower fuel costs, is certain to be widely adopted."[52] Honeywell's vice president claimed that in the house of tomorrow, each room would be heated to the desired temperature and this would "end family rows."[53] Taking up zone control as the next frontier of thermostat development, Honeywell pioneered a model whereby the temperature of different rooms would be independently controlled. The heterosexual, nuclear family would no longer simply be brought together through a single thermostat. Zone heating indicated the beginning of a shift away from evenness and toward the economic and social benefits of "varying house temperature."[54] At this moment, thermal differences and variations in the home were not only acknowledged but also turned into a potential source of profit. Everyone would now be free to pursue their own thermal needs. To do so, of course, they would need to purchase more thermostats.

Technologies of thermal variation were marketed as compensation for family differences as well as for the new architectures of the 1950s, including the ranch house with "more outer wall space" and "larger expanses of

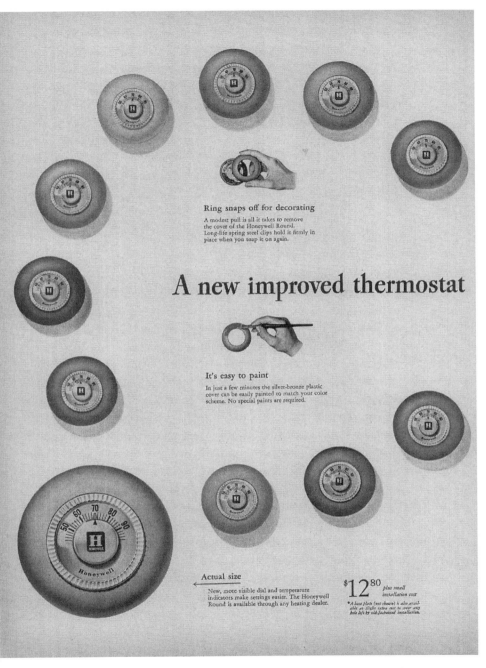

FIGURE 1.4. In an advertisement for the Honeywell T-86 thermostat, roundness and a multicolored palette (each thermostat is a different color) help to synchronize temporal, visual, and thermal registers. Honeywell, "A New Improved Thermostat," *Life*, November 9, 1953.

roof and window space."[55] They were part of a larger shift in the design of climate-controlled indoor environments, as Daniel Barber has documented.[56] The zone control system enables adjustment for "family living habits" as well as "for heating problems caused by picture windows, cold north winds, winter sunshine."[57] In 1955, Honeywell offered the Electronic Moduflow, an "entirely new temperature control *system*" consisting of an outdoor temperature "super-sensor" (designed by Henry Dreyfuss and integrating the new round design) that could remotely communicate with an indoor thermostat.[58] Honeywell argued that this "wonderful new comfort" was based on a new principle: "To have constant comfort you need varying—*not fixed*—indoor temperature."[59] Discomfort, Honeywell now claimed, came from *constant* indoor temperatures. In 1956, the president of Honeywell reflected that the company's "successful campaign on zone heating" had sold the public on more sophisticated and complex controls.[60]

The new thermal control products, even as their functions remained limited, capitalized on thermal differentiation. And although vernacular forms of thermoception had existed long before thermostats were invented and continued despite the deployment of homogenous temperatures, thermal differences started to appear in public discourse associated with thermostats. Technical articles began to refer to "individual comfort requirements."[61] A 1955 article in the *Baltimore Afro-American* begins, "Women of the world awaken. You've suffered long enough. Now is your chance to speak out and stand up for your rights: if your husband is a fresh-air fiend."[62] It goes on to describe the plight of the "long-suffering wife" who makes a "hasty dash to the furnace thermostat" and "prepares hot biscuits for breakfast, as an excuse to light the oven." The article is illustrated with an image of a husband jumping in the fresh air, and his wife shivering in the background (figure 1.5).[63] Such discourses, which draw from the language of both women's rights and the emergent civil rights movement, began to describe the thermostat setpoint as an individual preference with political importance.

In the years that followed, the conflict over the thermostat grew, turning into a "battle for the thermostat" that sold even more complex controls. In a 1964 article, one author speculates, "Many marriages might be saved—and many friendships, too—if science could come up with some sort of human thermostat that would enable everyone to be comfortable at the same room temperature." Wives, he observes, are usually too cold, and men—the "hot-blooded types"—can never win the war.[64] Writing a decade later, a wife laments her husband's preference for the cold and describes her own "subversive handling of the thermostat."[65] She warns her friends to bundle up when

FIGURE 1.5. An illustration reveals a woman shivering as her husband jumps in the cool air. From Ralph Mason, "If Your Husband Is a Fresh-Air Fiend," *Baltimore Afro-American*, December 3, 1955.

they visit. As these discourses intensified and consolidated the gendered tensions of the 1960s, they revealed that the heterosexual couple and the nuclear family were no longer single thermal units; each was a set of individuals who possessed diverging thermal desires. Gender conflict was temperature conflict. Almost all of these examples featured a woman subjected to her husband's thermal preferences, forced to occupy an environment that slowly sapped heat and energy away from her body. Feminist action involved seizing control over the thermostat.

Such conflicts escalated in the 1970s as the energy crisis prompted both a rise in heating oil prices and a series of thermostat regulations. Richard Nixon urged homeowners in 1973 to turn their thermostats down four degrees to "help get the country through an oil-short winter."[66] An op-ed describes one husband's "creative plan" for coping with the energy crisis: to keep the thermostat at 40°F and wear flannel long underwear. "Just think how pleased President Nixon will be!" the author writes.[67] He mocks his wife's reaction, describing her "insolence," "tantrum," and "bitterness" at the prospect of turning down the thermostat. Enduring cooler indoor temperatures in the winter came to embody a masculine, national pride. One reporter describes efforts taken by suburban Americans toward this end: "Their feet are freezing and their fingers blue, but they won't give in, won't turn on their furnace heat." It was the "greatest suburban shiver of all time," mitigated only by wood fires, Pak-A-Robes, and snug sacks and Hibernators, a "sort of sleeping bag with arm holes" that sold out in Sears stores.[68] Asking Congress for authority to mandate conservation measures in his 1979 "Crisis of Confidence" speech, Jimmy Carter argued that these low-tech energy-saving methods, including turning down the thermostat, were not simply common sense but acts of patriotism. To reset one's thermostat, to endure uncomfortable temperatures, and to be too hot or too cold—these thermal practices were recast as patriotic, environmentally conscious duties to save hundreds of thousands of barrels of oil a day.

Thermostat controls, long a concern of manufacturers, engineers, building operators, and occupants, became a focus of Carter's administration and an object of regulation. In 1977, in the wake of a cold wave that brought snow to Miami, negative temperatures to locations across the South, and a natural gas shortage across the country, Carter urged the American people to set their thermostats at 18°C/65°F during the day and 13°C/55°F at night. Two years later, in the summer of 1979, Carter signed a proclamation that limited thermostat settings in commercial buildings to a summer minimum of 26°C/78°F and a winter maximum of 18°C/65°F.[69] The following year he extended the program. Honeywell's executive vice president instructed the company's divisional vice presidents to find ways to conserve energy through controls, to transform conservation into "big business."[70] The company's microelectronic programmable thermostat was released in 1979, allowing users to set programs that save energy, money, and the environment. One consultant said in 1980, "The revolution in computers couldn't be better timed. It came with the energy crisis, and together they fell right in Honeywell's lap."[71] Digital thermostats emerged out of conservation.

Even after Ronald Reagan dismantled Carter's thermostat regulations and even with the new microelectronic programmable thermostats, the battle for the thermostat continued to play out in homes and in office buildings across the country. A self-identified "energy miser" recounted in the 1980s, "What a great feeling I had when turning down the thermostat—that surge of patriotic pride as the temperature dipped from the fuelish 70s to the sane 60s." But he was "antagonize[d]" by his wife, who "dressed as if she didn't have a cent to her name" as compensation for the low temperatures.[72] The energy crisis did not lessen the gendered conflict playing out at the thermostat, but it did offer a righteous platform for cold-inclined men. Another story describes a couple "'about to strangle each other' in a struggle for control of the thermostat." The husband argues, "We have to dial it down like everybody else."[73] As an energy-saving tactic, one man even decided to use only media—rolled-up newspapers—rather than logs to heat the home.[74] The normative thermoceptive regime in this moment was not one of evenness but one that lauded the capacity and will to endure discomfort and in turn ridiculed and feminized those who tried to keep warm.

The rhetoric of the energy crisis faded, but the tension between the "Colds" and the "Hots" has endured over the decades, with "trench warfare, full of sneak attacks, spies, loud recriminations, unwarranted accusations, and periodic battle cease-fires that usually don't hold very long."[75] One writer argues, "Compromise is difficult because, in general, the Colds [women] tend to be unreasonable in their demands."[76] The struggle over the thermostat occasionally resulted in violence. Such was the case when a man argued with his brother over the temperature in Illinois in 1991 and subsequently shot him as he attempted to unlock the thermostat.[77] Most frequently, thermostats became a means of sexual harassment. Male employers yelled at female employees for turning up the heat, refused to turn the heat up, put lockboxes on thermostats, and kept female employees from blocking cold-air vents (though the men did offer to keep the women warm themselves).[78] At the Blandin Paper Company in Minnesota, men turned up the thermostat in the women's restroom, making it uncomfortably hot.[79] A police officer increased the temperature at a convenience store in Vermont to 32°C/90°F, and when the female attendant went to turn it down, he assaulted her.[80] A man used a thermostat to blow up his home after being ordered to stay away in a divorce action.[81] And in 2018, after an investigation into the use of smart home devices, the New York Times described how the thermostat was being used as an instrument of domestic violence, with abusers suddenly increasing the heat or turning off air-conditioning systems.[82] In these examples, the thermostat became a direct means of communication, a convective and violent medium for control over the atmosphere in which others were immersed.

While early thermostat advertisements constructed an ideal thermostatic subject bound to the thermal unit of the heterosexual nuclear family, discourses and enactments of thermal conflict revealed the untenability and violence of this universal subject position. Repeatedly, these descriptions, published in newspapers over the latter half of the twentieth century, publicly articulated thermoception as a sense shaped by gender. These advanced a vernacular sense of thermoception, which, despite the advances in zone, variable, and programmable heating systems, revealed the ongoing impossibility and unwillingness to accommodate thermal difference.

STANDARDIZING THE SETPOINT

The deployment of the thermostat in American homes enabled people to manipulate temperature through the definition of a numerical setpoint. But in public spaces, industrial sites, and institutional buildings, the thermostat often remained invisible and inaccessible to inhabitants. A teacher, boss, building manager, or owner typically determined the broadcast temperature. While the battle for the thermostat in the home was a battle over who controlled the thermostat, in schools, workplaces, and other commercial sites it was a struggle over the definition of a standard temperature. In the development of such standards, the thermostatic subject emerged not only in advertising, as a user position, but also in scientific and engineering models, as an ideal thermal receiver.

In nineteenth-century heating systems, which were often advertised by salesmen and installed by contractors, there was rarely a standard setpoint. Even with the installation of thermostats, any given building could be set to any given temperature at any given time. The second president of the American Society of Heating and Ventilating Engineers (ASH&VE) lamented the lack of "scientific consideration" in these years: "Until about 1890 the business of heating and ventilating had been largely based on the most ancient rule known to engineers, the rule of thumb."[83] In an 1894 issue of *Heating and Ventilation*, an engineer voiced a similar concern: "The majority [of those working on heating and ventilation] are more anxious about getting work and money than about the mere art of heating."[84] With the creation of the American Society of Heating and Ventilating Engineers later that year, an imperative for standardization was articulated: "to establish a clearly defined minimum standard of heating and ventilation for all classes of buildings."[85]

The numerical definition of an optimal temperature for human comfort, one that could standardize thermal systems across buildings and geographies, emerged in the institutional deployment of ventilation and air-conditioning systems. As Gail Cooper describes in *Air-Conditioning America*, the technical HVAC community began "to take seriously the idea of creating a man-made indoor climate" with the advent of air-conditioning, one that could be defined in relation to a set of standard environmental attributes.[86] Although there were earlier proposals of optimum temperature levels, these standards emerged after a debate in the 1910s about the proper ventilation of buildings, especially schools.[87] On one side of the debate, fresh air advocates claimed that it was healthiest to allow the free flow of air through open windows. On the other side, proponents of air-conditioning argued that mechanical ventilation would create the optimal environment for learning. In 1913, the New York State Commission on Ventilation was organized to settle the issue. In appointing the commission, the governor pointed out the lack of scientific studies to determine "the most fundamental facts." It is not known, he pronounced, "what temperature should be maintained in public-school buildings," or even "whether a constant temperature or a varying temperature is more beneficial."[88] In pursuit of such thermal knowledge, the commission placed subjects in rooms with varied heat, humidity, air motion, and air composition; recorded their physiological condition; and tracked the "quantity and quality of intellectual product per unit of time."[89] Ultimately, to the dismay of those in favor of mechanical ventilation, the report came out in favor of fresh air.

At this moment, when the need for an artificial climate was itself under question, ASH&VE established a research laboratory at the US Bureau of Mines in Pittsburgh, Pennsylvania. Set up in part as a means to "bolster public confidence in the ventilating engineers' expertise," the research was driven by a need for "quantitative accuracy" that would not only help engineers design systems but supply them "with the surety of quantitative values in the rugged debate before the public in general and regulatory agencies in particular."[90] Among a range of experiments, the ASH&VE Research Laboratory tested theses on the "skin effects" of temperature, humidity, and air movement to determine the ideal climate.[91] In 1922, ASH&VE released the *Heating and Ventilating Guide*, a document that contained a "synthetic-air" chart that would later become known as the comfort zone chart.[92]

Laying a zone of comfort across a psychrometric chart, the comfort zone chart provided a set of definitive, numerical relationships between temperature, humidity, and human sensation (figure 1.6).[93] The chart features a series of curved lines across the grid of humidity and temperature: this was described as the "effective temperature," which indicated "one's feelings of warmth," a predecessor of today's "feels like" temperature.[94] In the center of the graph is mapped an ideal "comfort zone" with a precise "comfort line" drawn down the middle. The determination of the comfort line was based on research in the ASH&VE laboratory's psychrometric chambers, which included a three-hour test that recorded the thermal impressions of twelve subjects, largely personnel from the laboratory and from the US Public Health Service. It was also based on a two-hour test on five women and nine men, and although the group included a farmer, a mechanic, and a "laborer," it was largely composed of people employed in indoor white-collar jobs. A fifteen-minute study of one hundred subjects, largely students from the Carnegie Institute of Technology, supplemented these tests. In selecting the subjects who would define comfort, the researchers argued, "Great care was taken to get as nearly as possible a representative group of people."[95] But even with an attempt to include diversity of occupation and gender, these originating thermal subjects were predominantly white men employed (or seeking employment) in white-collar occupations. Moreover, as each subject was encouraged to "dress as he normally did," the existing formulations of gender and class were hard-coded into the comfort zone.[96] In an image of the representative subjects, for example, it appears that the majority of the men in the study wore suits and ties (figure 1.7). In contrast, in the data collected from the two-hour tests, all women were recorded as wearing either "light" or "extra light" clothing.

Although the charts were understood to apply to all bodies, the range of people whose opinions constituted the universal was limited. And even as the researchers realized that they could never fully account for all variables (especially metabolism), they embedded in the comfort zone chart a new form of thermal subject: a universal thermal receiver. While at the thermostat interface the thermal subject was being calibrated to sense as part of a family unit and to desire a "thermal monotony" that would ease household conflicts and uphold patriarchy, in the comfort zone graph, the receiving subject was imagined as a standardized thermal sensor, one whose body would react in the exact same way as any other to a given set of thermal stimuli. Created by researchers invested in the justification of mechanical ventilation and air-conditioning, it provided a rationale and structure for a

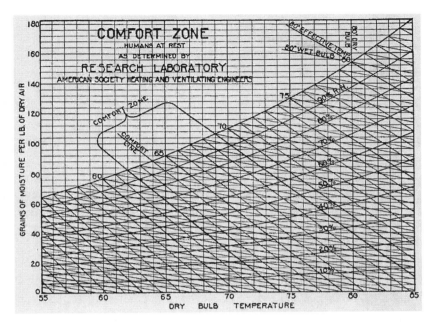

FIGURE 1.6. The comfort zone chart identifies the optimal temperature for human comfort. From American Society of Heating and Ventilation Engineers, *American Society of Heating and Ventilation Engineers Guide* (New York: 1924–25).

FIG. 1. REPRESENTATIVE GROUP OF 130 PEOPLE WHO SERVED AS SUBJECTS IN DETERMINING THE COMFORT ZONE

FIGURE 1.7. The original subjects of the comfort zone. From F. C. Houghten and C. P. Yagloglou, "Determination of the Comfort Zone," *ASHVE Transactions* 29 (1923): 362.

system of broadcast temperature, which, like the home heating system, distributed the optimal temperature to identically thermosensing bodies. In this process, these bodies became an imagined mass, a set of thermostatic recipients. As Daniel Barber says of the work of architects Victor Olgyay and Aladar Olgyay, who also developed a bioclimatic chart, the "comfort zone was above all a space of the *normal*."[97] Like the individual members of a family depicted in pre-1950s thermostat ads, thermoception in the comfort zone graph was imagined as uniform, regardless of race or gender (even as race and gender continued to be reified through climate determinism). The graph offered no space for differences among thermal subjects.

As the comfort zone chart was circulated to heating and cooling engineers, contractors, and building owners to guide decisions about system operation and to determine the correct setpoint for institutional spaces, it brought with it a quantitative definition of comfort and a universal thermoceptive body. In many ways, the comfort zone chart's thermostatic subject, even though it would continue to shape architectural design and heating and cooling systems into the future, is a historical anomaly, notably distinct from earlier ways of thinking about temperature. Humoral medical theory and environmental determinism had long assumed that people were shaped by their thermal environments. It had been more common to believe that bodies are affected by temperature, and that one's sense of temperature varies depending on one's body, than to believe that the body is an independent sensor. The comfort zone chart, in contrast, imagined the thermal subject as a blank slate. The sense of thermoception would stay unchanged even by thermal communications.

The comfort zone chart went on to have a long legacy, shaping research on temperature for the century to come. It was listed in military bibliographies and used to support soldiers' performance. It was used in factories to increase productivity. And it was revised, updated, and nuanced by the ASH&VE over the following decades. In 1966, the expanded and renamed American Society of Heating, Refrigerating, and Air-Conditioning Engineers (ASHRAE) released Standard 55, titled *Thermal Environmental Conditions for Human Occupancy*. ASHRAE 55 came to be regarded as "universally applicable across all building types, climate zones, and populations."[98] ISO Standard 7730—the international standard for "moderate" thermal environments—was later developed in parallel with a revised version of ASHRAE 55.[99] At the outset, Standard 55 described "thermal comfort" not simply as a physiological process but as a subjective experience: "the condition of mind that expresses satisfaction with the thermal environment and is assessed by subjective evaluation."[100] As in the comfort research from prior decades, thermal subjectivity was assumed to be shared and uniform.

From the 1960s onward—even as ASHRAE 55 defined a universal standard for human comfort—a more complex model of thermoception was developing, one that was tied to the data orientations, multiplication of variables, and computational potential in digital systems, and one that moved toward an acceptance of vernacular forms of thermal difference. In thermal comfort studies, a significant shift took place in the experiments of physiologist Povl Ole Fanger at the Technical University of Denmark. In 1970, Fanger published *Thermal Comfort*, which, like so many of the models that had come before, proposed that a person's perception of thermal comfort was dependent on environmental factors (including air temperature, air speed, relative humidity, and radiant temperature), their level of activity, and the thermal insulation of their clothing. Like many previous studies, including the earlier comfort research of the ASH&VE laboratory, Fanger placed subjects, wearing uniform clothing, in a climatic chamber. Like prior researchers, Fanger concluded that no major differences in thermal satisfaction and thermal sensation could be attributed to age, gender, location, body type, biorhythm, or time of year.

Where Fanger's work diverged from existing studies, however, was in the incorporation of a new and critical element: a rating scale for subjects' self-assessment of comfort. Thermal subjects were asked to describe the ambient conditions as cold, cool, slightly cool, neutral, slightly warm, warm, or hot. Based on this subjective feedback, Fanger proposed an equation of comfort that connected physiological and psychological parameters. *Thermal Comfort* offered a set of operationalizable and predictive variables that transformed thermal comfort into an equation, which today, as thermal researchers acknowledge, "lie[s] at the root of practically every set of thermal regulations in the West."[101] The index Fanger initiated is called the predicted mean vote (PMV) model, which can be overlaid onto a psychrometric chart in order to account for the thermal opinions of the people in any given environment. Fanger also established a second index for thermal difference: predicted percentage of dissatisfied (PPD). This is a way of indicating the percentage of people who feel thermally dissatisfied, even if the majority of participants believe the environment to be neutral or acceptable.

Fanger's work introduced a new approach to thermal communications to the scientific and technical communities. In these indexes, there was a fundamental acknowledgement of differences in thermoception, even given the exact same set of thermal conditions. They featured a variable comfort zone rather than a "comfort line." The PPD index predicted that there would always be a diverging thermal opinion. The PMV-PPD diagram could therefore be used differently from the early comfort zone chart—it granted much more room for variation across thermal environments and for localized adaptations. This model was subsequently incorporated into ASHRAE Standard 55 as the thermal sensation scale (see the chart). This enabled a calculation of how far a given climate deviates from the "optimum," defined as a sense of thermal neutrality for the greatest number of people. As the thermostat wars were raging, the oil crisis of the 1970s was putting pressure on heating bills, and the US government was setting standards for thermostats, thermal comfort research began to explore "the edges of comfort, searching for how cold or warm it could get before getting uncomfortable."[102]

The ASHRAE Thermal Sensation Scale

+3 hot
+2 warm
+1 slightly warm
 0 neutral
−1 slightly cool
−2 cool
−3 cold

In the decades that followed, Fanger's research shaped assumptions about "ideal" environments and standard setpoints. The 2013 version of ASHRAE Standard 55 contains two methods for determining acceptable thermal conditions. The "graphic comfort zone method" is a more nuanced model of the comfort zone chart, with several added variables for consideration, including clothing insulation, metabolic rate/activity, and radiant temperature. The second, a computer-based method, uses the PMV-PPD model to define the following acceptable limits of comfort: the PMV range must be less than +0.5 (at the warmest, it must be halfway between neutral and slightly warm) and greater than −0.5 (at the coolest, it must be halfway between neutral and slightly cool), with a maximum of 10 percent of the people dissatisfied. In other words, the dominant standards that determine the thermal operations of buildings across the United States and around the world continue to be guided by a standard thermal subject, even as the standard acknowledges thermal difference at the outset. (The 2013 edition of Standard 55 says, "Because there are large variations, both physiologically and psychologically, from person to person, it is difficult to satisfy everyone in a space."[103]) Fanger's PMV-PPD index provided a way around the need to address such variations through the designation of a set of normative and majority-based parameters.

In the 1990s, thermal comfort researchers and members of the heating and ventilating community began to critique the universality of these measures, prompted in part by the expansion of cooling systems across the Global South. A 1992 special issue of *Energy and Buildings* on the social and cultural aspects of cooling targeted the ASHRAE standards explicitly: "We find the prospect of cooling the anticipated growth in tropical offices and homes to ASHRAE standards ludicrous from a power supply perspective, and terrifying from an environmental perspective."[104] In the collection *Standards for Thermal Comfort: Indoor Air Temperature Standards for the 21st Century*, Ken Parsons observes that since thermal comfort is a condition of mind (it is a psychological state, not a physiological one), it is "influenced by individual differences in mood, personality, culture, and other individual, organisational and social factors."[105] While calling into question the philosophy and accuracy of existing standards, Parsons argues that methods for prediction "will never be perfect."[106] Research papers from this period question how thermal standards might change if deployed across the tropics, often pointing out the inadequacy of "Western-defined comfort norms."[107] "There is no one standard of human comfort," Lawrence Agbemabiese and a group of collaborators write. Individual variability was understood to be inevitable due to the human body's ability to adapt alongside its capacity to inherit or acquire thermal orientations "through ailments, physical training, diet, clothing habits, living habits . . . sex, race, age, and other physiological conditions."[108] The imagined thermostatic subject of engineers and researchers was questioned, and vernacular thermal knowledges were suddenly given consideration in the field of thermal comfort.

While these researchers continued to pursue thermal comfort as their goal, they expanded their sense of what comfort consisted of and, as a result, generated a new imagined thermal subject. One model from the 1990s was the adaptive approach to thermal comfort, which predicted that contextual factors and, critically, past thermal history played a role in people's expectations and preferences. In their 1998 proposal for the adaptive approach, Richard J. de Dear and Gail Schiller Brager criticized existing standards for prescribing "a narrow band of temperature to be applied uniformly through space and time" and, in particular, for viewing people as "passive recipients of thermal stimuli."[109] Opposed to this static model, their proposal echoed discourses of interactivity in the late 1990s mediascape, suggesting that people "play an active role in creating their own thermal preferences."[110] While some forms of adaptation were acknowledged in prior models (particularly the possibility of adapting behaviorally by changing clothes or turning up the heat), de Dear and Brager pointed out that there is also the possibility of intergenerational genetic adaptation, acclimatization over the course of a lifetime, and psychological adaptation, which they described as the "altered perception of, and reaction to, sensory information due to past experience and expectations."[111] After all, they commented, "personal comfort setpoints are far from thermostatic." [112] In turn, as another researcher articulated, "Discomfort is caused by excessive constraints being placed on these processes of choice and adjustment, rather than by temperature itself, except in extreme conditions."[113]

The shift to new models that could encompass a broad range of thermal preferences was grounded in a critique of earlier test conditions. Former studies were based on test results obtained in a climate chamber, and the applicability of these results to real contexts was "insistently and increasingly questioned."[114] The adaptive model acknowledged that in the real world there are "full adaptive possibilities[,] from operable windows through to task-ambient air-conditioning."[115] The selection of test subjects also started to be critiqued. While studies of temperature began with a "base in military research and studies of fit, young people,"[116] new research on air-conditioning started to be carried out in the field—in Thai office buildings, in Ghanaian homes, and in the factories of northern India—with a wide range of subjects.[117] Beyond this, researchers pointed out numerous variables that shape thermoception but remain absent from standards, including "time (exposure duration), time of day, time of year, adaptation, age, gender, mental activity, preference and past experience."[118]

These were calls for a multicultural model of temperature that could account for the diversity of thermal experiences, the global contexts in which thermal technologies were deployed, and the complexity of "comfort" as a social and psychological phenomenon. But they failed to gain traction in the universe of standards. It was not until 2004 that the ASHRAE standard was expanded to include the adaptive-comfort model, and even then, it was an optional approach. The adaptive subject remained at odds, at its core, with broadcast thermal infrastructures. Those who advocated for adaptive comfort imagined, in place of broadcast systems, a world where no one would be subject to unwanted broadcast transmissions, whether human or "natural." One researcher argued, "Individual control seems to provide the solution. How can someone complain if they control their own environment?"[119] Echoing this, another claimed: "Even with such modifications to thermal comfort evaluation, thermal comfort for all can only be achieved when occupants have effective control over their own thermal environment."[120] Equity, they hypothesized, could only be achieved with a new kind of infrastructure and mediation that bypassed standards and centralized systems altogether. It would be well into the twenty-first century before digital technology developers realized this project.

NEST AND THE RISE OF THE SMART THERMOSTAT

In 2011, a new thermostat was released on the market. Nest, a "smart" thermostat that learns about its users' thermal preferences and programs the environment to suit them, generated an explosion of news coverage and reviews not unlike Honeywell's T-86. The *New York Times* claimed in a headline: "Home Thermostats, Wallflowers No More."[121] The thermostat of decades prior was recast as an "ugly, beige tool," a "mass-produced hunk of plastic," and a "blah-looking controller."[122] Other tech start-ups, including Toon and tado°, released smart thermostats in the following year, capitalizing on Nest's popularity. Companies with long-standing investments in heating, cooling, and energy technology, including Honeywell, the Carrier Corporation, Schneider Electric, Ingersoll Rand, and Emerson, all developed competitors for the Nest. In 2014, smart thermostats became the most commonly purchased smart home device.[123] On April Fools' Day of that year, Virgin America announced that Nest thermostats would be installed on its aircraft, with features such as "destination acclimation," which would adjust the temperature to the climate of a traveler's destination. While the announcement was a joke, it was not far from the thermal realities of the future, in which personalized temperature control would promise to reorient the thermal landscape in the same way that digital media reoriented broadcasting: what was once a mass technology would now be instantly programmable, customizable, and oriented to individual users' preferences.

Nest was not the first smart, connected, or individualized thermostat. Ecobee, a Canadian start-up that was later backed by Amazon, had released a Wi-Fi-connected thermostat two years earlier. Honeywell had tested a smart thermostat decades before but never developed it due to "poor consumer response."[124] The Honeywell Prestige, which allowed users to do many of the same things as Nest, was available in 2011 but only to contractors who could install it as part of a heating or cooling system. Shortly after Nest's release, Honeywell filed a suit against the company, citing infringement on several patents, including a patent for controlling thermostats via the internet (the suit was settled four years later for an undisclosed amount). And despite Nest's emergence, Honeywell continued to dominate the thermostat market, even the sales of smart thermostats.[125] Moreover, the claim that these thermostats were the first to offer individual control was an overstatement: thermostats have long been controlled by a single person, and even with digital control, temperature is still manipulated by modulating an often-shared atmosphere. In short, Nest wasn't the first, the only, or the most widespread smart thermostat, but its design and popularity catalyzed a shift in thinking about the thermostat, from a utility system interface to a personal consumer electronic device. Instead of being seen as part of an architecture, thermostats were recast as an extension of the self.

This was in part due to Nest's development by former Apple engineers, including Tony Fadell, who designed the first generation of the Apple iPod. The interface of Nest, while echoing the circular dial of a classic Honeywell thermostat, is "a thermostat for the iPod generation"—there are no switches or mechanical buttons, only a dial reminiscent of an iPod.[126] The interface was initially "so enticing that people began to brag about their thermostat."[127] By allowing users to control their home's temperature remotely, through their phones and iPads, Nest connects heating and cooling systems to the existing landscape of mobile technology. In a subsequent release, the company added a motion sensor that can detect human presence and reset the temperature accordingly. Heating and cooling systems have long been used toward communicative ends, but the spread of Nest has encouraged users to manipulate these systems *as* a media platform and integrate them with existing media technologies. Like so many other apps released during this period, Nest allows users to monitor their environment, track usage, generate metabolic assessments, and construct a hyperpersonalized environment. This thermostat offers a fantasy of climate change in which constantly fluctuating temperature is a product of what Yuriko Furuhata describes as "atmospheric control."[128] Against the backdrop of oscillating and intensified weather conditions, such controllable microclimates have become more and more desirable.

At the same time as they enable their users to engage in controlled microclimatic change, smart thermostats are marketed as the forefront of sustainable design. A continually increasing number on Nest's website shows how many billions of kilowatt-hours thermostats have saved since 2011.[129] Ecobee allows users to "donate" their data to scientists working to improve the planet, boasting, "Sharing your data is a simple way for you to do something about climate change."[130] One of the selling points of these devices is that they are both environmentally *and* economically responsible. By modulating the energy used in heating and cooling, the smart thermostat promises to cut energy costs by lowering energy expenditure. It is this aspect of thermostat design that appeals to many digital developers and investors. Fadell, reflecting on the fact that around half of the average residential energy bill goes to heating and cooling, remarks, "That's a lot of money but everyone ignores the thing you control it with."[131] The thermostat is not simply a personal interface for one's thermal environment; it is a means of controlling the flow of money to utility companies.

The smart thermostat, like the "dumb" thermostats before it, pairs apparently contradictory fantasies: on the one hand, it offers the fantasy of increased thermal control, and on the other, it offers the fantasy of ceding control to the system itself. The Nest's AI-based algorithm monitors and tracks human preferences and uses these as a baseline to produce an optimal environment. In Nest's advertisements, the people squabbling over the temperature have disappeared. No one controls the temperature—the thermostat "programs itself. Then pays for itself."[132] For its users, the device offers a dream of personalized thermal care grounded in a long history of racialized and gendered labor: "Nest looks out for you."[133] The algorithm is articulated as caring simultaneously for the environment and for people's financial well-being by minimizing energy usage. At the same time, Nest is a "digital nudge technology," as Natasha Schüll describes, which means that it does not passively fulfill thermal desire but rather is an active, sensing, and guiding technology, "a sort of compass to help individuals navigate a world of choices."[134] Nest's interface features a green leaf that glows brighter the more its user turns the dial in a direction that requires less energy, nudging users in that direction. Although the Nest remains a home thermostat that works via convective media on an entire room at a time, it conceptualizes the home as a personalized thermal sphere, a bubble around a self and free from any memory of a family with diverse thermal needs. Like so many thermostats before, it rests on class distinctions—it's targeted primarily toward those who own their home, not renters. Cumulatively, the smart thermostat hails a thermal subject who can pursue self-interested desires (Carrier's smart-thermostat marketing tells consumers, "You're zero degrees of separation from the degree you want"), inhabit a personal thermal sphere, and engage in technological consumption to remedy the problems of climate change.[135]

During this same time period, another digitally inflected mode of thermal coordination gained traction: participatory thermal sensing. While the smart thermostat capitalizes on the desire for personalization generated by digital media, participatory thermal sensing builds on digital media's promises of democratization, updating Fanger's predicted mean vote method for the digital age. Universities have formed test beds for a number of these projects, including Thermovote, developed by researchers at the University of California, Merced; the ZonePAC system deployed at the University of California, San Diego; and the TherMOOstat app at the University of California, Davis. TherMOOstat, launched in 2015, allows staff, faculty, and students to provide feedback on the indoor climate to the university's facilities management team. Like the smart thermostat, the app is pitched as an energy-saving device and offers participatory thermal sensing as a way to stop the overheating or overcooling of personal space; it is described as better for both comfort and the environment. TherMOOstat generates crowdsourced comfort, replacing a preestablished setpoint with a collectively generated sense of temperature.[136]

Unlike smart thermostats in the home, participatory thermal-sensing projects still rely on a centralized authority that mediates requests. Crowd-Comfort, an app first pitched in 2013 for use in the workplace, enables its users to report problems, thermal or otherwise, to their building manager. CrowdComfort turns humans into the building's sensors, extending the perceptual capacity of those who manage it. Honeywell developed the Vector Occupant app, which enables users to report if they are too hot or too cold, again through a message to the building manager. In these systems, even as the thermal standard is contested, thermal control remains mediated by a central entity—facilities management—and depends on a central heating or cooling system.[137] Yet even as these offer a thermal ideology that appears to grant subjects agency over their environment, they are ultimately designed as a means of increasing employee productivity in workplaces.

These apps constitute the latest update of a thermal capitalism that modulates workplace temperature as a means of increasing productivity. Documenting the struggle over the introduction of office and factory air-conditioning, Gail Cooper recounts how the early air-conditioning industry worried that marketing technologies for employee comfort would run against the traditional values of hard work and productivity, and wouldn't appeal to American industry.[138] Carrier Corporation sought another rationale for its campaign, and eventually, as Cooper traces, "the industrial comfort campaign was abandoned."[139] In place of this, Carrier articulated a rationale of increased efficiency and productivity through industrial cooling. Predictably, the relationship between personal comfort and productivity "found more favor among managers of white-collar workers than it had among factory supervisors," and Carrier's campaign was aided by abnormally hot summers in 1949 and 1952, when some New York firms were forced to send employees home early.[140] In place of a standard human subject who is most productive at a set temperature, the digital temperature systems of the 2010s have offered an alternative that mobilizes the rhetoric and ideals of a digitized liberal democracy. All thermal preferences can be registered, afforded a vote, and used to determine the majority's preference, and changing the temperature accordingly is then represented as a form of thermal care, not by an algorithm, but by a company in service of its workers. But as Aaron Freedman rightly points out, control remains with management rather than with labor. As a counter, Freedman offers a rallying cry in the socialist magazine, *Jacobin*: "Workers must seize the means of cooling production for ourselves."[141]

Comfy, like the participatory thermal-sensing apps, is a smart-building technology that appears to fulfill this dream to relocate control from the building manager to the employees, harnessing the capacities of mobile technologies, networked heating and cooling systems, and the building occupants' desires for thermal comfort. Like the other projects, the smartphone app allows office occupants to send a message to the building to warm their space, cool their space, or just to let it know "I'm comfy." Pitching the app, the CEO of the Silicon Valley start-up argued that since the "number one complaint of office workers is being too hot or cold," Comfy would increase employee morale and productivity.[142] But what distinguishes Comfy is that the app's users, once connected to the building's heating and cooling systems, receive a "personalized thermal adjustment." Comfy extends a new mode of thermal communication to the white-collar workplace, one in which users are not polled for the ideal temperature in a consensus-based model but can each exist at their own preferred temperature. In other places, architects consider how a "placebo effect" might be achieved through dummy thermostats that could, as Gökçe Günel describes, "provide the illusion of control to [users] without compromising on the system's efficiency."[143]

Smart thermostats and thermostat apps collectively represent a shift away from the classic thermostatic subject. They draw inspiration from adaptive-comfort models, and their designers articulate these liberal, personalized technologies as a way to reconcile thermal differences and the thermostat war, whether it takes shape as a gender or a class war. In the promotional materials and discourses about these technologies, the new liberal thermal subject is not one who receives the ideal thermal message but one who appears to have the ability to choose their own temperature and is freed from the tyranny of others' thermal desires. However, the smart thermostat and participatory thermal sensing do not give their users thermal autonomy, the ability to determine the conditions of their own thermal being. They will not be subjects free and able to pursue their own thermal desires but will instead be subjects implicated in even broader systems of thermal management.

On the one hand, this occurs because, as they outsource the process of thermal management to algorithmic, machinic, and corporate control, users become dependent on technologies that inevitably break. In one particularly well-publicized episode, thousands of Nest thermostats turned off in January 2016.[144] Shortly after this, the Hive thermostat turned up the heat for users in the United Kingdom to almost 32°C/90°F. And in December 2017, some Nest thermostats stopped working during an extreme cold snap, a fault that the company blamed on the inability of users' HVAC systems to "share power effectively."[145] But even beyond the publicized outages and glitches, the failure of smart thermostats is evident in regular user complaints. One user reported on social media, "Below freezing outside and the device failed, allowing our home inside to drop to 50°F this morning."[146] Others recounted their thermostats going offline unexpectedly, changing the temperature unpredictably, or failing to register their presence or preferences. Even if, at times, the distinction between the algorithm predicting preferences and a glitch producing errors has been unclear, user responses have materialized as frustration with the failed promise of thermal care.

While smart-thermostat marketing has anthropomorphized the technology, imbuing it with agency and attentiveness to the user's thermal needs, ceding thermal control to a digital thermostat has gone beyond ceding agency to an algorithm. Nest connects its users to a network whose data is shared and can be used for remote control. This represents another key shift in thermopower. For companies such as Honeywell, this is an extension of long-standing practices of linking homes into networks through security systems, but for many smart thermostats, the networked nature of the device is driven less by consumer desires than by the potential to connect to large-scale utility companies. The devices finally enable utility companies to do something they have long sought to do but for which they have lacked the appropriate communications system: remotely alter the temperature of their users' homes.

This has been a goal of power companies due in part to the existing configuration of the electrical grid and to the difficulty of storing large amounts of electricity. Many users share temporal rhythms as well as responses to thermal conditions, which collectively result in peaks of electrical usage that strain an aging electrical grid, often when it is exceptionally hot or cold and there is not enough energy to meet everyone's thermal needs. For decades, energy companies have developed demand response programs as a solution to this problem. In agreements signed with individual households, energy companies have offered incentives for families to disconnect some heavy-usage appliances, such as water heaters and air conditioners, during critical peaks. The difficulty with these programs has been a communicative one: utility companies could send out a shutdown request, but customers could opt out or simply choose not to turn off appliances. The utility company would not actually be able to track whether the shutdown had occurred. The communication in such programs is one-way: from utility company to consumer. As a result, these programs have remained logistically heavy, difficult to enforce, and expensive for the industry, especially as the utility companies often have provided or installed thermostats or switches for users.

Smart thermostats—and users' enthusiasm for them as personal consumer devices—offer a new way out for utilities, one that enables them to reach into users' homes, alter the temperature, and better manage the flow of energy across their networks. Utility companies have begun to partner with smart-thermostat companies, incentivizing the purchase of smart thermostats not only through the combined marketing of personalized thermal comfort and environmental benefit but also through rebates and discounts for customers. Some utilities have even purchased start-ups that are developing smart thermostats (Hive, for example, is owned by British Gas). Nest created Rush Hour Rewards, a platform for utility companies to launch demand response programs, and EnergyHub developed a platform interface to connect thermostat companies and utilities. In 2013, Southern California Edison offered the peak time rebate (PTR) demand response program in conjunction with Nest. In 2018, Nest announced that it would provide a million thermostats to low-income homes as part of its Power Project.[147]

Nest found a new way to describe what it offers: with the ability to manipulate actual thermal conditions, it is not just reducing the load on the grid for utilities; it is supplying *power*. Following the closure of the Aliso

Canyon natural gas reservoir in Los Angeles, which occurred after a massive leak in 2015, Nest was positioned to offer a "virtual power plant" to help make up the difference.[148] The company made a deal with Southern California Edison to "deliver" around fifty megawatts—but this supply would be achieved by the reduction of people's usage, through Nest's control of home temperatures. One news report argues, "Nest's new deal could be seen as a replacement for a power plant, at much lower cost."[149] These programs are framed as payments for not using energy, a way to "empower" customers to control their bills, and a personalized alternative to mass setbacks or power outages. Despite such rhetoric, these deals quantify people's "personalized" thermal spheres and allow them to be exchanged between energy companies and tech companies, independent of exterior thermal conditions. As one author writes, "Nest's revenue from utility companies outweighs revenue from the sale of those thermostats. . . . Nest is making more by supplying your data—and your tacit control—to the big energy companies."[150] This is another form of thermal capitalism: the manipulation of people's thermal spheres in order to allow for the reallocation of power. And as heat waves impact both users and utilities, they catalyze an expansion of demand response programs: the more extreme the weather, the more justified the smart thermostat appears.

Moreover, as artificial intelligence is incorporated into thermostat networks, in both homes and commercial buildings, temperature is made subject to algorithmic control. Decisions about people's thermal exposures are less frequently made by individual or corporate thermal "caretakers"; they are increasingly the output of complex software systems. Whether in homes enrolled in peak demand response programs, in workplaces with AI-enabled thermostats, or in buildings that deploy massive sensor networks to keep track of environmental conditions, temperature is an algorithmically mediated refraction of data on thermal preference, past thermal responses (including moments when users choose to override the system), energy costs, and the assessment of occupant presence and activity. The last of these is often gauged by another thermal medium, infrared imaging (discussed in chapter 5). Mitsubishi's 3D i-See sensor, as one example, is able to locate occupants based on their heat signature, and the cooled air can then be pointed directly at recipients.

Like so many other digital technologies, smart thermostats offer a fantasy of increased agency while exposing the user, the family, and workers to new forms of vulnerability. The most obvious danger might seem to be from malicious hackers. Take, for example, the case of British Gas's Hive, which had to be changed in thousands of homes after "it was exposed as a 'burglar's dream.'"[151] Or when two hackers demonstrated that they could load ransomware onto a smart thermostat and then demand that the thermostat's owner pay up or be subjected to 37°C/99°F or no heat at all.[152] The idea of being held hostage by one's technology is a familiar one, but it masks the fact that users already turn over their vital capacities to technology and utility companies, and these companies in turn allocate decision-making to an algorithm. For example, per Nest's terms of service, users waive the right to trial by jury, the right to participate in a class-action lawsuit, and the right to claim punitive damages—no matter what temperature the company subjects them to.[153] A focus on malicious hackers also masks the ways that thermostats continue to be a means of enacting control and violence within the home. In the case of domestic abuse, for example, thermostats are already used to hold people thermally hostage.

What this history of digital temperature control highlights is that even prior to the thermostat war of 2015, when thermal standards were publicly called into question, untethered from a normative male subject, and reanchored in logics of eco-economic efficiency, an array of consumer technologies had emerged to capitalize on thermal difference. These materialized the dream of adaptive-comfort researchers from prior decades, but they explicitly took shape in the neoliberal, personalized, mobile, and digitized landscape of the 2010s. In this period, it became possible for a person to drag a finger across a screen and subtly shift the thermal landscape around them. It became possible for many to conceptualize themselves as part of a thermal democracy, in which all of the inhabitants of one's environment were equally entitled to register a digital vote about the ideal temperature. And it became possible for electrical companies and technology companies to instantaneously manipulate the thermal landscape of millions of people at once.

NETWORKED THERMOPOWER

Connecting these moments in the history of the thermostat, this chapter reveals a cross section of the ways that thermopower operates not only through infrastructures for heating and cooling but in system interfaces, in thermal standards, and through the visual, aural, and tactile forms that help people to make sense of temperature. Thermopower exists in the definition of normative practices of temperature control, whether these situate thermostat use as a duty of the housewife or as a means of saving the environment. Thermopower extends through the creation of standards and regulations, as well as through cultural practices of heating and cooling. Although thermal technologies are often represented as neutral infrastructures that simply maintain human life, this chapter shows how heating and cooling function as material substrates for communication. It reveals how thermostats, standards, and the discourses about them calibrate bodies as perceiving subjects. As they standardize such operations and crystallize political forms, thermal media remain one of the least visible and most effective means of managing people and behavior.

Alongside the transition from mass media to personalized, networked, and digital media, there has been a parallel transformation of thermal communications. The mass thermostatic subject emerged as an ideal in the early twentieth-century heating and cooling industry in the United States. The comfort zone chart and the ASHRAE standards formalized a language of classical thermal communication that was then exported around the world. This vision, as Daniel Barber writes, "was an ode to stability and to comfort as a specific form of economic exclusion."[154] The adaptive-comfort models, even as they began to emerge in the late twentieth century, didn't take hold as the basis for a substantial technological project until the 2010s, when the existing mediascape could support a critical shift in thermal management. If in the twentieth century thermal care was described as a responsibility of a wife or the state (as ASHRAE and other standards were adopted in United States federal regulations), the turn to acknowledging thermal difference and the emergence of personalized thermal technologies shifted responsibility for temperature to the individual. For a sweating white-collar man in thermostat promotional videos, this transfer offered empowerment and upward mobility. At the same time, it created an environment that, in providing the so-called solution to thermal discomfort, could legitimate decreased funding for heating assistance programs and require instead the adoption of smart thermostat technology. In a neoliberal thermal sphere, thermal difference was portrayed as neither structural nor conditioned— temperature is a preference and an economic privilege.

New technologies of temperature control today are dominated by a digital and personalized ethos, offering the possibility, developers claim, of finally controlling one's thermal environment, granting a sense of atmospheric control.[155] The promise of these technologies appears to be one of thermal autonomy, but it is an autonomy in which one's body is no longer a subject of unwanted broadcasts, no longer at war with differently thermosensing beings, no longer responsible for the thermal care of others, and free from standards and regulations. In place of this, the subject is imagined, in the vein of adaptive-comfort research, as always able to change the temperature. People are apparently freed from the will of traditional thermal mediators who once determined and altered the setpoints: maids, janitors, building operators, corporations, husbands, and mothers. And in the process, these technologies promise what, retrospectively, seems to have always been the failure of a prior thermostat model: the separation of temperature from energy. These technologies appear to offer true thermal freedom.

And yet, temperature remains intimately connected to energy systems. And in these technologies, bodies are connected to a vast network of thermal exchange. Far from becoming autonomous, the inside of one's home now shifts with the economic models of utility companies, and one's bodily temperature depends on access to charging stations and the ability of manufacturers to keep wages low. Thermopower is not granted to the individual; it is dispersed across the network.

This expansion in thermal control, in the capacity to precisely modulate the rhythms of heat and cold, occurs against a backdrop of increasing public recognition of climate change. The fluctuating climate legitimates the expansion of these technologies by creating in people's bodies an expectation, a desire, and a technical solution: An expectation that the climate will shift, unexpectedly and dramatically, and that it will be possible for such shifts to be a product of human agency. A desire to retake that agency and to feel in control of climate changes. A technical solution that provides a fantasy of atmospheric control, one that is not simply imagined but is dreamed in one's skin. In turn, people's bodies are calibrated with the expectation of this control, of personalized and yet externalized thermal care, even as the system relocates responsibility to them. Now they must be adaptive.

2 COLDSPLOITATION THE THERMAL ATTRACTIONS OF COOL AIR

The best temperature is no temperature at all. Many heating system engineers, textile designers, and building architects share this assumption. In thermal standards and best practices, the optimal state is "neutral." Thermal comfort itself is "the condition of *minimal stimulation* of the skin's heat sensors and of the heat-sensing portion of the brain."[1] In this paradigm, the optimal thermoception is nonperception. The ideal thermal message is one in which nothing is received. In such an approach, perception mechanistically echoes thermodynamics. Thermal subjects want to be neither hot nor cold. Since thermoception registers an energetic exchange between bodies and their environments, thermal comfort's ideal subjects simply desire a world in balance with their thermal selves.

The Goldilocks theory of thermoception, in which everyone wants to be "just right," ignores many other forms of thermal desire. It fails to account for why people actively seek out experiences of hot and cold. Visitors to Sweden's Icehotel choose to spend the night in below-freezing temperatures (−5°C/23°F). Yoga practitioners perform sets of bodily postures in hot environments (40°C/104°F). People frequent saunas, partake in polar bear plunges, and pursue thermal affects. In everyday life, some forgo the thermal homogeneity of central heating and let cold breezes into hot buildings. On warm days, some people bask in the sun, collecting radiant heat. Thermal desire cannot be reduced to a quest for neutrality any more than sexual desire can be reduced to the pursuit of cold women by hot men.

While thermal comfort science has historically ignored the desire for thermal affects, philosophers, architects, and historians regularly comment on temperature's appeal. Comparing the warmth of a flame to sexual activity, Gaston Bachelard claims that the "gentle heat" of "man made warm by man" lies at the source of the consciousness of happiness.[2] Inspired by people's desire for heat and cold, Lisa Heschong calls on architects to design experiences of "thermal delight."[3] There is an "aesthetics of heat," Boon Lay Ong shows, in experiences like drinking a warm beverage or taking a warm bath.[4] Some scholars have even prompted thermal comfort research itself to move from "thermal boredom" to "thermal pleasure."[5]

This chapter is about thermal desires and delights. Alongside psychological processes, architectural influences, and thermoregulatory dynamics, I describe how senses of heat and cold are cultivated by and manifest as *thermal attractions*. Just as media generate visual and aural attractions, so do thermal media, from refrigerators to fans, stimulate and channel thermal affects. And as radio, cinema, television, and the internet harness peoples' attraction to images and sounds, they also circulate thermal experiences for profit. Thermal capitalism not only organizes temperature to produce more efficient bodies, it also traffics thermal attractions as objects of value and exchange. These attractions are not generated in a vacuum. Like all other forms of perception and communication, they emerge within a matrix of cultural ideologies and practices. They offer pleasure and entertainment, all while situating subjects within broader structures of thermopower.

One of the most often debated thermal attractions, from the nineteenth century to today, has been the production of a sense of cold. For well over a century, cultural critics have asked why many people, especially in the United States, desire the cold. For some authors, the development of cooling systems is merely a battle in the struggle of humankind to escape "the slow torture of hot weather."[6] Others unpack the symbolic layers of cooling: ice is a symbol of the United States, Tom Shachtman argues, tracing the desire for cold back to the nineteenth century.[7] Even in 1855, a *DeBow's Review* article pronounces, "Ice is an American institution—the use of it an American luxury—the abuse of it an American failing."[8] Thermal comfort researchers ask, "Why do Americans so passionately wish to be cold in summer?" For many, the answer exists in technology: iceboxes, refrigerators, and air conditioners enable a sense of mastery over nature and weather, itself a long-standing American preoccupation. After all, the technology and trade of ice production, distribution, and consumption were dominated by the United States.[9] In *Refrigeration Nation*, Jonathan Rees argues that "Americans' devotion to refrigeration helped define them as American," in part because the United States "took to refrigeration of all kinds faster and more extensively than the rest of the world."[10]

These "hedonic pleasures of the cold" do not emerge out of a universal thermal need or simple exposure to cold technologies, Hi'ilei Julia Kawehipuaakahaopulani Hobart points out—they are "buttressed by conditioned performances of colonial legacies."[11] In "Cooling the Tropics," Hobart documents how the use of ice in Hawai'i is intimately connected to a settler colonial project that offers "sensory pleasure as a measure of global power."[12] The spread of ice machines and ice cream in the nineteenth and twentieth centuries, Hobart shows, instilled a sensorial ideology. The consumption of coldness became a way to "civilize" and "Americanize" Indigenous Hawaiians. Hobart's account is just one example of how nineteenth-century ice— shipped around the world from Boston or manufactured by ice machines— cooled drinks and other commodities in colonial outposts, from claret wines in Calcutta to champagne in Havana.[13] An 1892 article of *Ice and Refrigeration* states, "Only in the cities through which the never-ending procession of tourists passes is ice to be found."[14] The desire for cold did not simply stem from a desire to master nature; it was also deeply connected to colonial expansion.[15]

This process—in which a sense of cold reproduces colonial and racial ideologies—was part of a larger process of thermal colonization. Natural and artificial ice were also leveraged as a medium of thermal colonization in a second way. Used to preserve food in transportation, ice facilitated the construction of a massive, global cold chain. Like the sale of thermostats and cold sensations, this was a project of thermal capitalism. Manipulating temperatures allowed for the preservation and then sale of products at the optimal moment. Commodities could be sent to a wider variety of markets, across much greater distances, and this transformed food consumption and production.[16] Perishable goods from around the world were made accessible in the West. Ice underpinned the construction of global food systems and enhanced capacities for settler colonial occupation and trade.

The distribution of ice and the sense of cold materially enabled the expansion of colonization and affectively conveyed it as a thermoceptive regime.

Thermal colonization was enacted not only through affective and logistical means but also existed in the deployment of thermal knowledge as the legitimation of colonial projects. As many historians have shown, climate determinism and racialized thermal knowledge were integral to colonial expansion and to white supremacy. In the United States specifically, the sense of cold was bound up in the depiction of Arctic and Antarctic environments. As polar explorers rushed north in the nineteenth and twentieth centuries, their expeditions—and their encounters with coldness—were major media events covered in the print news, captured in photographs and shared as lantern slides, recounted in illustrated lectures and published journals, and eventually documented in films. Reading explorers' reports, poets, novelists, and artists imagined the emptiness of the landscape in terms of temperature. These representations, Hester Blum observes, emphasized "the North's frigid stillness."[17] The image of the Arctic as a frozen wasteland cast it as a terra nullius and served to "blank out" the landscape of Indigenous peoples. Thermal colonization encompassed the creation of sensations that could justify expansion, dispossession, and settler inhabitation. Even the scientific knowledge of climate change constituted through ice core samples, Jen Rose Smith reveals, was shaped by ice and ice geographies' "inextricable relations" to ongoing coloniality.[18]

Thermal knowledges were used to scaffold colonial projects and to sell cooling technologies. The spread of air-conditioning in the twentieth century, Marsha Ackermann documents, was explicitly legitimated by colonial and white supremacist ideologies. Environmental determinists, including Ellsworth Huntington, C. E. Winslow, and S. Colum Gilfillan, developed justifications for mechanical cooling, especially the idea that inhabiting the cold was evidence of racial superiority.[19] In *Cool Comfort: America's Romance with Air-Conditioning*, Ackermann shows that the adoption of air-conditioning was not immediate. Only after coolness could be aligned with health, progress, civilization, and whiteness was it embraced by an urban middle class. The Carrier Corporation, along with other air-conditioning manufacturers, also used racist depictions of Indigenous people to sell cooling technologies. Whether conveyed by ice or air conditioner, coldness has often communicated a superiority to racialized others.

This chapter does not track a genealogy of cooling technologies—the authors mentioned above have produced excellent works in this area. Rather, I tell a story about how cold attractions, as they emerged within a landscape of national identity and colonial power, also took shape in a media landscape, both as they were installed in spaces of media distribution and as they were explicitly crafted as forms of bodily communication. I focus on a formative moment at the beginning of mass cooling: the period from the 1910s to the 1930s in the United States, when a new set of cold attractions began to hail media consumers to take part in new thermal pleasures. These attractions pioneered what I call atmospheric coldsploitation—a mode of affective mediation that simultaneously exploited the sense of cold through colonial and racist representations of the Arctic and Antarctic, new forms of advertising and publicity, and emerging technologies of cooling.

Prior to this moment, in the nineteenth and early twentieth centuries, for people that consumed it, cold was largely a discrete, material commodity. It was picked up, transported, harvested, and then consumed. Sold directly to consumers as a commodity, it was packed into iceboxes or offered as cubes in beverages. Cold, stored in the medium of ice, was encountered not unlike the coal that was purchased from a coal dealer or the images and sounds that were circulated via photographs, newspapers, phonographs, and optical devices. It was often an object of personal consumption that worked directly on the body and could be carried in one's hands. Ice was always "a mobile phase of matter," Rafico Ruiz, Paula Schönach, and Rob Shields observe.[20] Moreover, ice also had a finite temporality. Like the newspaper, it was an ephemeral medium. In its use and representation, ice often had a transformative, magical potential: to alter objects and change their physical composition. It was contained, applied, and used as an instrument.

From the late 1910s to the early 1930s, coldsploitation media—including films, radio, and air-conditioning systems—turned from the discrete possibilities of ice to the ambient and the architectural. This shift in orientation, from cold objects to cold air, paralleled a broader shift from media objects to media environments. Coldsploitation media were enabled by the development of motion picture theaters, architectural exploitation and new advertising practices, and the ascendance of radio. Instead of being purchased as a commodity, the cold was broadcast to a mass audience via the manipulation of air and airwaves. In turn, cool air became a visceral attraction to be consumed alongside visual and tactile media. By the end of this period, a new set of thermal media—air conditioners—also helped to reshape cold as something to be inhabited. As a new technology, air conditioners became a cold attraction in themselves, but they did not precipitate on their own the massive thermocultural shift to convective media, the techniques of atmospheric manipulation used to redirect heat and cold. Coldsploitation films acclimated audiences to the idea of being chilled through the theater air well before cooling technologies were installed. These thermal visions mobilize what Laura Marks calls "haptic visuality," a form of visual mediation that elicits an embodied, tactile response.[21] And although such projects exemplify what Yuriko Furuhata describes as "thermostatic desire," coldsploitation viewers were calibrated to seek thermal attractions rather than programmed thermostasis.[22] They were oriented less to environmental control than to encounters with thermal otherness.

The cycle of coldsploitation reveals that thermal media have not always— or even predominantly—been motivated by a movement toward thermal nonsensation. Rather, this case shows how technologies of thermal desire consolidate and stimulate interest in hot and cold experiences and channel these into the pursuit of thermal attractions. These technologies are enlisted in projects of thermal colonization and capitalism. Coldsploitation reinforces a refrain throughout this book: bodies are not simply determined by climate. Thermal communications are not abstract ideas. They are part of the material patterning of people's thermal exposures, delivering affects that shape associations with such exposures and layering histories of colonization into one's sense and inhabitation of the environment. Thermopower operates both through the creation of normative subject positions, charts, and standards (as demonstrated in the previous chapter) and through the production of affective landscapes.

Revisiting this history today is critical because, as I discuss in this chapter's final section, in the early twenty-first century a new mode of cold attractions developed. While convective media work via the manipulation of cool air (and are enabled by atmospheric control), new forms of conductive media transmitted cold attractions through contact. Smartwatches, smart sheets, and smart clothing all leverage digital technologies and the process of thermoelectric cooling, used in microprocessor operation, to redistribute thermal sensations. Here, too, thermal media matter. These technologies are often marketed as a return to personal temperature control, hearkening back to ice or even fans, but like the smart thermostat, they do not grant thermal autonomy. Instead, they hook users into a vast grid of electrical supply. They interconnect with databases and make bodies subject to algorithmic manipulations. They require maintenance, upgrades, batteries, and a massive network of environmentally damaging digital technologies. And they ask users to take responsibility for satisfying their own thermal desires, even as these desires are cultivated by an expansive landscape of thermal attractions.

Although this turn to personal cooling purports connections to ice, it is a new development in the lineage of cold attractions. Rather than indexing an environment or a shared condition, temperature here is a property of an individual body. Through personal cooling, one's body can be divorced from the environment, apparently made invincible to threats of convection, whether from climate change or particalized contamination. In these technologies, the consumption of cold is not an attunement to thermal affects through sight, sounds, environment, or atmosphere. Rather, it involves hooking into a digital circuit that conducts cold directly to the body. This coincides with a broader shift in the perception of icy spaces that Cymene Howe observes, from "natural configurations, as landscapes, as patrimony, as mystical spaces or as that which can both provide life and bring death" to entities where "we see heat absorbed: the atmosphere enacting a thermodynamic play."[23] Although conductive cooling technologies are still coded in relation to Arctic and Antarctic environments, they no longer depict atmospheric excursions across lines of thermal difference, an exposure of the white, Western body to the "natural environment" of the racial or ethnic other. Instead, thermal difference is encoded as an internal state that can be controlled completely, achieved without travel, and manipulated through a thermostatic interface.

COLDSPLOITATION AND CONVECTIVE MEDIA

In the 1920s, the media landscape of the United States was changing. The first radio broadcasting stations in the country were established at the start of the decade, and by 1923, there were over five hundred stations. Modern advertising took shape not only in radio but also in a growing number of print media and magazines. A web of telephone and telegraph services extended farther and farther across the country. Cinema's audience doubled: from forty million weekly viewers in 1922 to eighty million by 1928.[24] People's bodies were coming into contact with media in new ways—and this reshaped how heat was communicated. Media technologies of prior decades were often held and touched. Heat was conducted to hands from newspapers hot off the press. Wax cylinders were placed into machines, and listeners worried if the cylinders' expansion and contraction in the heat and cold would affect the quality of the music. Moving images could be viewed in a single-user kinetoscope or in a cramped five-cent theater. People could feel the heat of the projector and other audience members. Many theaters, adapted from storefronts and other existing architectures, also amplified environmental heat.

With the expansion of mass media, communications were increasingly ambient and atmospheric. Radios transmitted voices and music through the spectrum, invisibly directing beams from centralized towers to permeate homes and bodies. A wave of movie theater construction resulted in the proliferation of downtown cinemas, and advertising "exploitations" extended the content of the screen to lobbies and sidewalks. This was also the era of the movie palace, the massive purpose-built architectures for viewing films, complete with costumed in-house ushers and elaborate lighting effects. Fans and ventilation systems helped to reduce the sense of the thermal presence of others—as well as the cinematic mechanism itself—in the theater. A century before twenty-first-century "atmospheric media," these new media forms harnessed the environment—architecture, airwaves, and air itself—as part of their communication.[25]

Although wireless telegraphy and cinema had been in operation since the end of the nineteenth century, these media became more ambient during this period—part of a broader turn to the atmospheric. "The discovery of the environment," Peter Sloterdijk writes, took "place in the trenches of World War I."[26] As gas warfare was introduced into battle, the target shifted from the enemy to the environment that permitted the enemy's survival. The atmosphere and air, "as media," Sloterdijk shows, became the object of explicit consideration, monitoring, and manipulation, not only in science but also in culture and aesthetics.[27] This occurred in the wake of both gas attacks and airborne reconnaissance. Aerial perception and warfare extended the battlefield into the atmosphere, and in the process, "the supply of images become the equivalent of an ammunitions supply."[28] Climate took on a new role. Rain, snow, fog, and temperature, because they conditioned when planes could move, affected the supply of ammunition images in new ways.

It was not only the atmosphere that was discovered in this moment— the cold was also increasingly an object of explicit consideration. During World War I, journalists reported on the unbearable cold, troops standing in frozen slush in the trenches, deaths by hypothermia, ice blocking supply lines, the elaborate logistics of delivering underclothes and blankets, and the visual spectacle of snow falling on the front lines.[29] Climatologists speculated about war's atmospheres and the connections between war and winter. One remarked, "The fact that this war is being fought in the winter means hundreds of thousands of dollars to the manufacture of winter supplies in the United States."[30] Rudyard Kipling chronicled the battles fought between Italian and Austro-Hungarian troops in the Dolomites and its frigid temperatures, ice tunnels, and snow trenches. In the "white war," as it was later called, the Austrian Corps of Engineers dug an entire ice city into a glacier. Cold waves struck Europe during this time, affecting the troops. But the cold also hit the United States. The winters of 1916–17 and 1917–18 set record low temperatures along the US East Coast, and blizzards took out communication and transportation infrastructures. In New York City, an "army of 9,500 'snow fighters'" was marshaled to help clear the streets.[31] At the same time, the 1918 influenza pandemic generated another set of meanings for the cold. Although distinguished from the "common cold," the experience of thermal variation—sudden chills or fever—signaled the onset of influenza. Those who were sick shuddered and huddled under blankets. Thermal manipulations, from the prescription of aspirin to the administration of hydrotherapy, were used in treatments. Speculations about the flu's climatic ties, like the war's climatic ties, circulated through public discourse.

Following the war and the pandemic, a new wave of explorers visited snowy, icy, and cold environments. In 1921, the first expedition to climb Mount Everest was launched, and in the following years, George Mallory embarked on repeated attempts to summit the mountain until his final climb in 1924. Chronicling the expeditions, journalists focused on "battles" with wind, ice, and cold, and observed that exploration was often limited not by altitude but by temperature and atmosphere. This was the final stage of the "heroic period" of polar exploration: Antarctic explorers continued to launch expeditions, which were portrayed in the news as heroic struggles against the cold, even as their public lectures were losing popularity (figure 2.1).[32] And it was the dawn of a technological period of polar exploration, in which new aerial forms, including planes and radio systems, became critical to expeditions. The Arctic communications of Richard E. Byrd, Roald Amundsen, and Donald MacMillan, among others, afforded immense possibilities for cold narration. Cold, illustrated with snow, ice, and mountains, was increasingly sensationalized and the fight against it was militarized. One correspondent described the Amundsen expedition thus: "Explorers picture polar waste as [a] cold and hostile expanse of frozen water, fashioned by nature into never-ending phalanx of pointed weapons."[33] And in popular culture, cold was relayed through expanding media systems, from Tony Jackson's ragtime hit "Don't Leave Me in the Ice and Snow" (1917) to the popular radio show *The Clicquot Club Eskimos*, first broadcast in 1923.

In the atmospheric media of this period, a new thermal attraction and sensorial form emerged in the United States, especially in movie theaters: coldsploitation. In coldsploitation films, sometimes referred to as "snow

16 Months in Antarctic Cold Rivals Stories of Heroism on War Front

FIGURE 2.1. An image of Ernest Shackleton's Antarctic expedition published in the *Washington Post* features men "struggling through a polar blizzard," as the accompanying article says. From "16 Months in Antarctic Cold Rivals Stories of Heroism on War Front," *Washington Post*, December 3, 1916.

pictures" or "snow films," cold environments were a primary part of the attraction, if not the motivation for the entire film. During this period, Robert Flaherty's canonical *Nanook of the North* (1922) was released. Scott MacKenzie and Anna Westerståhl Stenport argue that this film did "more than any other film to codify how the Arctic was seen and imagined cinematographically well into the very last part of the twentieth century."[34] *Nanook of the North*, they show, came to stand in not only for the North but also for the cold itself. But *Nanook of the North* was neither the first representation of a cold environment nor the first coldsploitation film. There had been northern and polar films since the beginning of cinema, from films focused on the Klondike gold rush in the late nineteenth century, to Frederick Cook's *The Truth about the North Pole* (1911) and Georges Méliès's *The Conquest of the Pole* (1912), to travelogues such as *Alaska: Wonders in Motion* (1916).

In the late 1910s and early 1920s, however, Arctic, "Canadian Northwoods," and Alaskan films became immensely popular. In 1921 alone, northern films included *Nan of the North*, *Heart of the North*, *The Blizzard*, *The Call of the North*, *Flower of the North*, *Cameron of the Royal Mounted*, *The Golden Snare*, *Blind Hearts*, and *Snowblind*, among many others. Describing *The Iron Trail* (1921), a reviewer writes, "It is just the sort of picture that can be depended on to entertain nine out of every ten patrons, for these melodramas of the frozen North, when well produced, as in the present instance, always seem to make a distinct hit with the movie fans."[35] These are exploitation films not because they feature any illicit content—although the classical exploitation film featuring forbidden topics also emerged as a distinct category of motion picture during this time.[36] Rather, the coldsploitation film harnessed northern, polar, and cold environments as sites of sensationalism and spectacle, capitalizing specifically on their snow and ice.

By the time *Nanook of the North* was released in 1922, critics were already overwhelmed by the snow picture. One writes, "We have had snow stories, tales of the wintry north woods and Northwest Mounted dramas times without number."[37] Coldsploitation permeated nearly all genres. There were the snow comedies, from *The Frozen North* (1922) and *Cold Feet* (1922) to an episode of *Our Gang* (1926). Snow romances trapped lovers together in snowbound cabins. *The Man from Beyond* (1922) released Harry Houdini from frozen ice. *The Woman Conquers* (1922) was a "story of a society girl in the Arctics." Canadian Northwoods dramas—such as *The Man from Hell's River* (1922), *Jan of the Big Snows* (1922), *The Storm* (1922), *Where the North Begins* (1923), *The Eternal Struggle* (1923), and *Shadows of the North* (1923)—were exceptionally popular, as were Alaskan stories, including *Brawn of the North* (1922), *The Spoilers* (1923), and *The Alaskan* (1924). Filming on-site was a huge draw: films from *The Chechahcos* (1924) to *Santa Claus* (1925) were advertised as "actually made in Alaska."[38] And of course, the cycle also included the films of polar explorers, who, after losing popularity on the lecture circuit, looked to film to regain an audience; their films ranged from *South* (1920) and *Below the Antarctic Circle* (1922) to *The Great White Silence* (1924).

Exhibitors and critics collapsed the distinctions between the snow pictures' disparate ecologies and geographic locations. In these films, snow and ice were not only the primary feature of the environment but an attraction in their own right. Sometimes they were described as characters: in *Over the Border* (1922), "snow scenes, including a blizzard, feature [in] the picture."[39] Snow effects were written into scripts: in *Judgment of the Storm* (1924), a special note to the art department reads, "In shooting the foregoing title and all others that appear during the storm sequence, double expose drifting snow into titles." Snow was enlisted in physical gags: in *The Frozen North* (1922), Buster Keaton repeatedly slides on, falls into, and is submerged under snow. Critics and audiences homed in on the impressive visual displays of snow and ice, lauding them as exceptional spectacles, especially in dramas. In *The Heart of the North* (1921), it was the blinding snowstorm that gave the film "a true Northwest atmosphere."[40] A reviewer of *Shadows of Conscience* (1921) reports that the "blizzard scene is one of the most realistic ever seen on the screen. It fairly makes you shiver."[41] Audiences wondered at the climax to D. W. Griffith's *Way Down East* (1920), which involved the rescue of a heroine from a river's ice floes: one reviewer boasts that it surpasses the suspense and power of the gathering of the clans in *The Birth of a Nation* (1915).[42]

Snow and ice were exploited even in films not set in the cold. The elaborate snow and ice carnival in *The Young Diana* (1922), for example, was hailed as an achievement in stagecraft.[43] Films shot in polar environments were praised for the "realism" of their environmental representations. One reviewer writes of *Below the Antarctic Circle* (which was actually shot in Antarctica), "The most realistic of all the pictures, however, is that of the blizzard."[44] Reviewers of *Nanook of the North* claim that it was "the snow and ice which provide the spark that gives life to the drama"[45] and that the film was powerful because it was the "only picture feature ever photographed wholly in Arctic conditions."[46] These expedition and filmed-on-location movies were directly compared to other movies as snow pictures: "If one thinks there were thrills in the ice-jam in *Way Down East*, . . . then this graphic story [*Nanook*] will undeniably hold one spellbound."[47]

Although snow and wind, as with many atmospheric phenomena, are typically "overlooked" in visual media because they are "a materiality with little visuality," during this period, air and cold became objects of attraction.[48] While audiences marveled at snow on screen, they also read extensive backstories about snow and ice in fan magazines. Snow films even had their own celebrities. One article describes these celebrities as snow men and women "who are always associated, in the minds of the public, with scenes of snow and ice."[49] In "Snow Stuff," the *Picturegoer* recounts the lengths to which filmmakers went to capture real snow: just as the shoot for *Back to God's Country* (1919) began, leading man Ronald Byron died from exposure, and the whole company got frostbite.[50] During the filming of *The Wheel* (1923), the leading lady fell two hundred feet down a mountainside. Hundreds of tons of snow and rock were sent hurling for an avalanche in *Wolves of the North* (1921). Truckee, California, was "overflowing" with production companies seeking snow scenes.[51] But for those who went to Yosemite, one article tells readers, the comfortable hotel and tourist restaurants meant that "snow scenes and luxury go hand in hand."[52]

When filmmakers couldn't achieve real snow, either in Alaska or in California, they faked it, with salt thrown in the air and a wind machine (one picture used over forty tons of salt for a winter scene). "It is often better and cheaper to fake a snow scene in this way," a journalist reports, "for the real stuff is generally too uncertain to rely on."[53] *Motion Picture Magazine*, showing a convincing image of a fake snow scene, challenges its readers to discern the difference between it and a real one (figure 2.2). Sometimes the scenes ended up being "half-real, half-artificial"—reporters imply that film crews themselves produce real weather with wind machines "used to create the whirling storms of winter."[54] But faking it also risked bad reviews. *Variety* slams *The Grip of the Yukon* (1928): "The snow was so synthetic it failed in its only reason. A wind machine can't cool the public."[55]

Even in the early 1920s, the motion picture industry was already producing films that reflected on the creation of atmospheric effects. In *Souls for Sale* (1923), a film about the movie industry, the picture's climax hinges on the intrusion of "real" weather on film production. A title card tells the audience, "The weather is always playing havoc with movie plans, and on the final shots a real hurricane threatened to wreck the artificial storm." Inside a big top tent, an audience of extras watches a circus scene unfolding. A film crew stands by with a wind machine (the very kind used to generate snow for the snow pictures), ready to create an artificial storm. But when the tent is struck by lightning, it starts to burn and the actual storm intrudes on the film set. "Keep grinding till it gets too hot for the cameras," the filmmaker yells as the circus audience, fleeing the burning big top, is subjected to sideways rain and fog clouds. In the film's final moments, the wind machine itself is used as a weapon. This is a typical Hollywood depiction of wind, as Hunter Vaughan observes: it is "evoked symbolically as something beyond human control and yet ironically is harnessed and manufactured."[56] But these scenes do more than simply reveal the follies of atmospheric recreation. They stage a fictional audience for whom real and fake weather are interchanged, whose bodies are overpowered by the spectacle. The collapse of the big top under the storm also evokes the danger of the snow: audiences of that period might have been reminded that only a year prior, the Knickerbocker Theatre in Washington, DC, had collapsed under the weight of a snowstorm, killing nearly one hundred moviegoers.

While snow, ice, and wind were exploited as images, intended to draw audiences in to a spectacle, "exploitation" at this time referred to more than just a kind of film. It described a particular kind of advertising and publicity. Theater owners and managers regularly traded information about the successful "exploitations" of individual movies, including their use of elaborate exterior displays, product tie-ins, and theater decorations. In previous decades, film promotion involved first banners and handbills and later ploys and staged scenes outside of theaters. But around 1914, the exploitation era

You
Never
Can
Tell!

When you see Zena Keefe in this scene of a Selznick picture, you'll probably hope that she doesn't contract a cold or something even more serious. But have no fear—the picture above tells the tale . . . of the painted snow-covered peaks; the fans blowing an electrical gale, and pounds of artificial snow. The exteriors actually were photographed in the frozen North. Not so the cabin scenes. Which goes to prove . . . You never can tell!

FIGURE 2.2. *Motion Picture Magazine* challenges its readers to tell the difference between real and artificial snow. From Emerson and Loos, "Building the Scenario," *Motion Picture Magazine*, April 1921, 41.

arrived, and it lasted through the late 1920s. *Exploitation* became an industry term, and the large production companies developed their own exploitation services.[57] In tandem with growing theater construction across the United States, exploiters sought to expand cinema's audiences. They looked for content that would not only draw in audiences by itself but also encourage the creation of atmospheres in theaters, lobbies, and on the street. Films were being assessed in terms of their atmospheric potential.

For this, snow pictures were ideal. Across the United States, exhibitors constructed elaborate theater environments of "snow" and "ice." This was certainly the case for *Nanook of the North*—exhibitors reflected that snow and ice "have been used to good advantage in nearly all theaters that have played the film"[58] (figure 2.3)—but even prior to its release, elaborate facades of snow and ice (in reality, paint, cotton, compo-board, and fake icicles) had been used regularly to publicize films such as *The Man Trackers* (1921), *The Idol of the North* (1921), and *Playthings of Destiny* (1921).

Snow film exploitation extended beyond what exhibitors called "cool lobbyology" to the alteration of the inside of the theater.[59] A Kansas theater manager went to great lengths to create an "atmospheric stage setting"— with a white backdrop indicating "snowy wastes" and bordered by trees touched with artificial snow.[60] To theater owners and operators, the actual location of the picture (Alaska, Canada, Antarctica) did not matter—facades were simply described as "arctic appearance" or an "arctic front."[61] New technologies were developed to enhance the snow image. The Brenopticon could superimpose light effects over the film, including an aurora borealis effect for pictures of the Far North or a snow effect for use during snowstorm scenes.[62] A Westinghouse theater switchboard allowed "light artists" to mix "cold, bleak colors for the tale of the frozen north."[63] Through such techniques, coldsploitation moved beyond the screen to the elaborate cultivation of a surrounding "cold" environment.

Some exploitations even experimented with actual snow and ice. Advertising *Nanook of the North*, the manager of a Niagara Falls theater placed blocks of ice containing "frozen little Eskimo dolls" in front of the area's downtown stores, alongside information about the film.[64] To get a "sufficiently chilly atmosphere" for *The Broken Silence* (1922), a theater manager froze photographic stills in a huge cake of ice.[65] While they were on display in the theater, as soon as one began to melt, the manager placed another cake in its place. In the evening, he sent these back to the ice plant to be refrozen. This stunt extended beyond cinema: in Seattle, records of "Any Ice Today, Lady?" (1926) (a song that asks "Why you don't order some Eskimo

FIGURE 2.3. Exploitations of *Nanook of the North* (1922) rely heavily on fake snow and real ice. From "Striking the Keynote," *Exhibitor's Trade Review*, September 9, 1922, 987.

water?") were frozen in ice and displayed in the University Music Store's window.[66] And of course, in the tradition of circulating ice water and ice cream, audiences of *Nanook of the North* were encouraged to eat Eskimo Pies, chocolate-covered ice cream bars, in conjunction with the film.

Through the atmospheres of coldsploitation run long-standing tropes of thermal colonization. Regardless of whether the films were set in the Arctic, Canada, or Alaska, these landscapes were made interchangeable in the depiction of the frozen North, and their geopolitical histories and colonial occupations were erased. The snow and ice created a setting for men and women to fall in love. It was a landscape for the assertion of moral order, where white heroes catch nonwhite "criminals." It was a set of elements—a character— to be battled. Some of these films completely evacuated the snowscape of all inhabitants save "savage" timber wolves. When Indigenous people did appear, they were typically reduced to racist stereotypes, often referred to as Eskimos and portrayed as villains attempting to thwart the well-meaning explorations of the white heroes and their police dogs. Coldsploitation was not simply the exploitation of the image of snow and sense of cold; it was the direct exploitation of Inuit labor, lands, and visuality. Coldsploitation generated "settler atmospherics," as Kristen Simmons describes, an environment in which "atmosphere becomes not only a medium for violence and control, but also one through which affects to demean are engineered."[67]

The trope of the frozen wasteland was critical to these imaginations. The spectacle of the snowy landscape was described as "trackless wastes and brooding silence" (*Where the North Begins*). Film director Charles Reisner, commenting on *Nanook of the North*, lauds Robert J. Flaherty for picking "a cold, dreary, dismal locale. He showed us that his people and their modes of living and the great joy they got out of almost nothing." Reisner suggests that this is a great lesson for audiences "who think [they] are not getting the best of it."[68] The image of the frozen wasteland was a means of coding the Arctic as empty, not simply through blankness but through coldness, which, as Blum notes, is "an ideological move analogous to that of early Europeans in the Americas."[69] Temperature was used in both the pictures themselves and the paratextual material to portray these landscapes as an uninhabited terrain, naturalizing an imperative for both the white characters and the United States to assert possession over Indigenous lands. At the same time, they relayed a sense of "temperate-normativity," which as Jen Rose Smith argues, cast Indigenous people who inhabited cold landscapes as both exceptional and aberrant.[70] The point of filming thermal environments, as Reisner writes, was not to communicate comfort but to give white audiences a sense of the superiority of their own thermal landscape.

In coldsploitation films, the white landscape offered a space for the projection of colonial and racial fantasy, and in the imaginations of the exhibitors—and evident in their many publications and displays—the white screen itself was directly connected to the blankness of the snow. One exhibitor designed a New Jersey exploitation for *The Storm* (1922) that depicts a snowy landscape in a screenlike square framed by snow-covered rocks (figure 2.4).[71] The complexity of a screen is described in terms of snow: "Everyone who has taken a walk on a sunshiny day, when the ground was covered with fresh fallen snow, can recall the glaring reflection . . . recall how it almost blinded you. That is exactly what happens when the light from a projection machine strikes a screen surface which fails to match it perfectly in color."[72] In another article, a critic uses snow as a metaphor to make theater exhibition problems tangible: if you sit in the first five rows, you walk out of the picture with "all the sensations of snowblindness."[73] To see a snow film was to have one's senses overtaken by whiteness.

With its images of snow, fake-icicle adorned lobbies, blizzard light projections, and melting ice cakes, coldsploitation was seen as a way to increase attendance and profit, offering "opportunities for displays of exceptional appeal in the summer."[74] Describing the showing of *Snowbound* (1921) in a hot Atlanta September, *Exhibitor's Trade Review* credits the exploitation

FIGURE 2.4. An exploitation of *The Storm* (1922) reveals the frozen north in a screenlike square. From "'The Storm' as Campaign Inspiration," *Exhibitor's Trade Review*, September 23, 1922, 1133.

with drawing additional business: outside the theater, "dripping icicles gave [a] cool inviting appearance and attracted [an] immense amount of attention."[75] Summer exploitation itself was nothing new. Even in 1915, Epes Sargent's book *Picture Theater Advertising* had already chronicled an array of tactics to combat the summer heat. It suggests: Replace velvet hangings with muslin and replace warm colors with white. Set up fans in the box office to lightly blow the hair of the "box office woman" or use them strategically in the lobby. Offer free ice water. Put up icicles meant for Christmas trees. Wool can be used for snow, but it will probably get dusty. Signage is key, the book claims—try these: "Trips to the North Pole, ten cents. Apply inside." "Cold waves ten cents each." "This is the snow storm center." "Cool as a cave or a cake of ice."[76] And although the book doesn't mention snow pictures, by the following year, an exchange manager observes that there was already a "clamor for snow, ice, north pole pictures during the hot months."[77]

Exploitation films, from the 1920s through the current day, are known for their direct, affective, and bodily communication. Coldsploitation was no different—the exploitation of words, images, and architecture was intended to have actual cooling effects. The snowy scenes of *Out of the Silent North* (1922) were "cooling and should prove to be a welcome summer attraction."[78] As Kansas City audiences entered *The Idol of the North* (1921), the display of a snowbound log cabin was imagined as having a "'cooling effect' on downtown crowds, sweltering from a hot sun."[79] An exchange manager proclaims, "A person simply cannot sit in a comfortable theater, gaze upon a snow scene and feel hot."[80] And in the fall of 1922, *Exhibitor's Trade Review* predicts that in the coming winter, "the records should show a decided decrease in theatre expenditures for white paint and compo-board,

An arctic front was used effectively at the Rivoli, Columbia, S. C., not only to give the summer sufferers some relief, but to boost "Playthings of Destiny," which contains a realistic blizzard scene.

FIGURE 2.5. The "arctic front" for *Playthings of Destiny* gives "summer sufferers some relief." From "Voice of the Box Office," *Exhibitor's Trade Review*, September 17, 1921, 1126.

due to the greatly shortened orders of exhibitors who want to do the atmospheric thing by Pathe's 'Nanook of the North.'"[81] This "atmospheric thing" here was, as in World War I, a way of targeting a subject indirectly through thermal media that manipulate their environment as well as their body's sensory capacities.

These exhibitors offered their audiences a haptic and synesthetic thermal vision. They believed that seeing cold would have a thermoceptive impact even in the absence of air-conditioning systems or ventilation. For them, cold media were not simply ice but words, images, and sounds. By 1922, when *Nanook of the North* was released, one author writes that already "advertising— copy that establishes in the public mind the coolness and comforts of the interior of the theatre on the hottest day—has implanted itself so firmly that it is being recognized almost as essential as the name of the attraction itself."[82] But alongside the words, theater managers describe the need for subdued lighting and "restful appointments of the interior."[83] Cold media could be used strategically, one suggests: "The blazing asphalt pavements, regarded as the seasonal enemy of good business, is really an asset when the theatre front is arranged to form a pleasing contrast."[84]

These techniques of marketing coldness were not entirely new. In theaters, ice water and ice cream had long been used as cold attractions. In 1907, an ice cream parlor owner modified his shop so that people could purchase an ice cream to consume at the movie theater in his store, while a set of fans lowered the temperature to 18°C/65°F.[85] "Ice cream parties" continued to be used in cinemas, even for non-Arctic themed movies, through the 1920s.[86] Synesthetic cooling—via image, sound, and text—was also a well-established tactic beyond the theater. For example, one preacher's use of "cooling hymns" during a hot summer in 1910 offered a form of "mental refrigeration" for his congregation.[87] And since at least the latter part of the nineteenth century, ice had been regularly used as an attraction in ice palaces, ice carnivals, world's fairs, and ice skating displays—often connecting to an imagination of the Arctic.

Coldsploitation extended and intensified these techniques, but this cycle of media moved beyond simply imagining that synesthetic vision or listening could cool. In this moment, harnessed to atmospheric forms, coldsploitation modulated the environment around a viewer to generate a set of thermal, bodily affects. Coldness was not only something that could be consumed, through ice or ice cream, to transmit thermal affects to an individual body. It was achieved through convective media, an atmospheric, architectural, and ambient mode of communication. While this came to the fore in the 1920s most powerfully in the rapidly growing film industry, coldsploitation was used across media platforms. In 1923, as New York City theaters featured elaborate ice displays, Vorbach Bros., a phonograph dealer, constructed a similar window display for their store, with fake snow, evergreen trees, and a sign that invited people to "enter the store and be as cool as the scene depicted in the window."[88] They credited the display with helping to sell at least ten phonographs. But in cinematic coldsploitation, managers took bodily cooling a step further. Given that, in many of these films, the "camera literally acts as a tourist" and "the pleasure in the film [was] in this surrogate of looking," coldsploitation afforded audiences the capacity to be not simply visual tourists but thermal tourists—to achieve pleasure in the sensation of thermal Otherness.[89]

As these films offered an Other thermal environment, they enlisted Indigenous people to provide the racialized labor of cooling audiences. Alongside the long history of class- and race-based divisions in the infrastructural labor of heating and cooling, the racial division of labor also surfaced in coldsploitation media in more direct ways. One white theater manager in the US South advertised that any of his patrons who felt too warm in the theater "would immediately have two colored attendants at his disposal to fan him."[90] On one occasion, when a patron called the manager, within a few minutes two Black men arrived, "vigorously applying a palm fan; and within five minutes more the ponderous individual called them off again."[91] *Picture Theater Advertising* suggests that theaters "dress a small black boy in Oriental costume and provide him with a feather fan. Let him stand or sit beside a sign that reads: 'If you can find a seat inside that is not reached by a fan, we'll send the boy in to fan you.'"[92] Racism and racial labor was enlisted in the project of cooling many white audiences. Moreover, whiteness and racial identification was seen by some as rivaling the power of summer heat in itself. One article in *Exhibitor's Trade Review* reports an "emergency action" during a heat wave, in which a theater director resorted to "appealing to pride of race" to offset the heat. The director decided that "nothing short of an Irish play would lure Brooklyn home-owners off their verandas."[93]

This sense of thermal Otherness was also key to the transmission of cold via radio. During the height of the coldsploitation film, from 1920 to 1924, radio moved from a new medium to an ambient presence. And it was during this period that many of the medium's patterns and practices were set in place, one of which was the audio travelogue. With messages carried from different climatic zones, radio proved an excellent way for audiences to encounter the cold, for producers to exploit the cold as content, and for station owners to exploit Arctic expeditions for radio's expansion. In the early 1920s, radio listeners began to hear sounds from and about the Arctic. Donald B. MacMillan's Arctic expedition charted not only the poles but the possibility of radio in northern environments. For MacMillan's expedition, Eugene MacDonald, one of the founders of Zenith Radio Company, provided MacMillan with shortwave radios.[94] On the expedition, explorers used the radio to broadcast regular updates as well as an elaborately staged Inuit concert in Etah, Greenland.[95]

These broadcasts did more than just carry information about a cold North; they articulated radio's capacity to bring different thermal zones together. Here, too, the work of marketing and publicity was critical in extending the medium's perceived reach. Zenith Radio's advertisements, which highlighted the MacMillan expedition, foregrounded the thermal ramifications of radio: "Inside the Arctic Circle, nine degrees from the North Pole, a little 89-foot schooner is frozen fast in the ice of Smith Sound. Abroad this schooner a group of brave men are enduring, as best they can, the desperate cold of the Arctic—cold that often drops to 60 degrees below zero. Human atoms in a boundless field of ice!"[96] Radio, the advertisement claims, enables Arctic listeners to break the oppression of the Arctic solitude and cold—through listening to, in this case, concerts from a warm location: Honolulu. The power of radio was not predicated on simply breaking down geographic distance but on bridging thermal contexts—bringing a sense of warmth to the cold and bringing a sense of cold to the warm, all via the airwaves. One 1923 article in *Radio World* explicitly appeals to the possibility of bridging temperature across time and space: "To amateurs, sitting in nice, cozy, warm rooms," the author writes, "imagine, if you will, that instead of being where you are, you are up in 'the farthest north,' where the wind is biting cold."[97]

As MacMillan's expedition revealed, and as Arctic filmmakers had known for some time, polar explorations were acclimatization opportunities that benefited not only colonial explorers but also their media. These explorations were the testing ground for the development of new cold-weather media technologies. Working out how to make media function in the cold expanded the reach of its visual and sonic regimes and, in turn, its capacity to transmit colonial ideologies. Describing explorer Richard Byrd's use of radio, a Kolster radio representative says, "This scientific expedition of fifty-five men will be more completely equipped with radio apparatus than any band of explorers in history. . . . It is a matter of profound satisfaction to us that Commander Byrd and his radio experts have chosen Kolster equipment. It has been selected to withstand the extreme conditions of 150-mile gales and temperature changes from 125 degrees above while crossing the equator to 50 degrees below in the Antarctic."[98] Similarly, capitalizing on Robert Flaherty's use of their portable projector and electric plant, an advertisement from J. H. Hallberg boasts, "The plant had to stand unusually hard knocks during transport to the North and back to civilization again."[99] Coldsploitation facilitated cold-weather productions' capacity to profit from the development of new media technologies, whether radio or cinema.

During this same period, popular radio shows also capitalized on the colonial imagination of the cold Arctic. *The Clicquot Club Eskimos*, which began broadcasting in 1923, offered musical performances of a "banjo orchestra" alongside skits such as an "Eskimo Chamber of Commerce" meeting in "Iglooville."[100] The program is notably sponsored by a ginger ale company. The announcer began one show by saying, "Look out for the falling snow, for it's all mixed up with a lot of ginger, sparkle, and pep."[101] Cold affects, transmitted by stories from the North, were connected to a cold product.

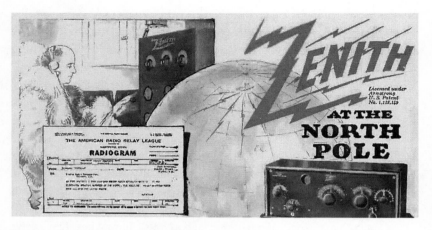

FIGURE 2.6. An advertisement for the Zenith radio describes how radio enables cold men in the Arctic to listen to concerts from Honolulu. Zenith, "Zenith at the North Pole," *Radio Age* 1 (1924): 52.

In the studio program, the group was dressed in parkas and the stage was accented with icicles, using classic coldsploitation imagery. One description reports that they "had suitable arctic scenery and dog-team sound effects to help them sell—not frosted ice-cream pies—but ice-cold ginger ale."[102] Another program report quoted a listener: "These programs make it feel thirsty!"[103] *The Clicquot Club Eskimos* moved across media and was featured in films and movie promotion, and its songs provided the soundtrack to at least one coldsploitation stunt: when a phonograph dealer converted an ice cream freezer into a phonograph and from it played the show's recording of "I Scream, You Scream, We All Scream for Ice Cream."[104]

While radio programs did not engage coldsploitation to the same degree as movie theaters of the time, they nonetheless attempted to exploit radio, as a set of real-time spectral transmissions, to relay a sense of immediate immersion in the cold and to create a sense of cold that could attach to other products. As cinema was exploiting the thermal synesthesia of vision, these transmissions attempted to exploit the thermal synesthesia of distant sounds. Collectively, these media evidence a shift in thermocultures. They harnessed the colonial imagination of the Arctic and of cold Canadian and United States landscapes as a means of thermal attraction. They leveraged new media forms to immerse spectators and listeners in a thermal landscape. In these cultures of thermal media, a new form of thermopower materialized. This was not a set of practices in which temperature (the cold) was held in discrete commodities (ice), transported, and individually consumed. Instead, mass media was used to generate, convey, and profit from thermal attractions. While thermostat discourses of this period were crafting an ideal, minimally stimulated thermal subject, coldsploitation was eliciting and cultivating thermal affects and desires. During this time, it capitalized on the fear of atmosphere following World War I and the influenza pandemic, portraying the environment as a dangerous and uninhabitable landscape, one in which the boundaries between human-created and "natural" atmospheres were unclear. At the same time, it mediated the racial conflicts, xenophobia, and Arctic exploration of the decade. But it did so by representing these landscapes as a spectacle of whiteness, a landscape that white audiences could consume as a thermal Other and that could, as these many exploitations suggest, help to reaffirm their sense of ownership over and security in their "own" thermal environments.

THE COMING OF COOL AIR

Cinematic coldsploitation extended beyond the Arctic images on screen and in lobby displays. From the 1910s onward, theater owners and managers also began to exploit the actual circulation of cool air. Alongside their investments in snow pictures and ice cakes, they invested in new cooling systems. At first, they installed electric fans and elaborate ventilation systems. But neither the meaning nor the impact of cool air could be assured. Cooling systems, too, had to be exploited. The Typhoon Fan Company offered a booklet of potential exploitations and urges owners to treat "the exploitation of the cooling and ventilating system in the same manner that producing companies treat the exploitation of their pictures."[105] They borrowed explicitly from the aesthetics of coldsploitation imagery with a ready-made slide for theater owners announcing, "Typhoon breezes cool and ventilate this theater" (figure 2.7). The company also included many ideas for lobby displays, including a giant thermometer that contrasted the temperature outside (hot) with the temperature inside (cool). In his 1915 publicity guide, Epes Sargent recommends, "No matter how perfect your ventilating system, use some small fans, if for no other reason than the moral effect."[106] Coolness and cold infrastructure had to be made visible. Exploitation made them part of the attraction. As was true for much coldsploitation of this time, the Arctic was enlisted in selling cold technologies. Chas L. Kiewert Co. boasts of their fans, "In the frozen North they don't need them. You do."[107]

FIGURE 2.7. The Typhoon Fan Company constructs materials to exploit their cool air. "Exploiting Your Cooling System," *Exhibitor's Trade Review*, October 19, 1921, 1534.

The ready made slide

But ventilation systems also promised something more than cool air. They promised *clean* air, and especially a freedom from odors and impurities. One ad for Sirocco fans told theater owners, "Banish the stagnant, stuffy air with its depressing odors." Their air-washing equipment offered "an ample supply of cool air from which all impurities are removed."[108] Cold air was able to intensify the cinematic attraction by lessening odors and heat from other bodies, which might remind people that they were indeed part of an audience. But cold, clean air, like the colonial fantasies of Arctic exploration, promised a blank sensory environment free from the heat and contamination of nonwhite or immigrant Others.

Narratives from this period more explicitly revealed the latent fears of what cool air might be hiding, preserving, or sustaining. H. P. Lovecraft's 1928 short story "Cool Air," published in the magazine *Tales of Magic and Mystery* at the same time as reports from polar explorations and descriptions of cold storage were being printed in newspapers and melodramas of the frozen North were being projected in theaters, begins: "You ask me to explain why I am afraid of a draught of cool air; why I shiver more than others upon entering a cold room, and seem nauseated and repelled when the chill of evening creeps through the heat of a mild autumn day. There are those who say I respond to cold as others do to a bad odour, and I am the last to deny the impression."[109] This is a story of cold's bad affects, of the narrator's encounter with his Celtiberian neighbor in a New York apartment building. Entering Dr. Munoz's door on a hot June day, the narrator is greeted with a rush of cool air but feels nothing but aversion: the chilliness "was abnormal on so hot a day, and the abnormal always excites aversion, distrust, and fear."[110] Befriending the ill doctor, the narrator learns of an elaborate cooling system, kept at 12°C/55°F–13°C/56°F, that alleviates the doctor's ailments. One day, the system breaks, and despite all of the narrator's attempts to repair the refrigeration, the doctor perishes in the heat—leaving only a note that he had died eighteen years prior. In Lovecraft's story, artificial refrigeration, long used to extend the life of food, extends the life of the dead.

Only a few years later, Lovecraft wrote another story, set in Antarctica, that again cast cool air as the origin of bad affects. The narrator in "At the Mountains of Madness" recounts a horrific polar expedition, the discovery of an ancient alien civilization, and the uprising of the long-oppressed shoggoths to destroy their makers as well as the expedition crew. Throughout the story, the narrator focuses on the terrifying Antarctic wind. Over the course of the narrative, the wind gains agency: it howls, mangles, ravages, and is "violent beyond anything we had so far encountered."[111] The shoggoths—created by aliens, repressed as slave laborers—are frighteningly amorphous forms, depicted on the cover of the magazine where the story appeared as an atmospheric, aerial, and wind-like entity (figure 2.8). They disrupt all other forms of communication—radio, planes, and speech. Both the shoggoths and the wind are racialized and "savage," and they wreak havoc on the scientists and the alien civilization.

In these two Lovecraft stories, which are examples of coldsploitation fiction, cool air is the habitat, the weapon, and the analogue of racialized Others. These narratives speak to the fear that accompanies cool air's promise to reduce other sensory stimuli and its offer of a pure and white landscape—a fear that cool air simply masks, and speaks to, the presence of Others.

It was during this period—when snow films were playing in cinemas, stories were being written about cool air, and radio sets were broadcasting voices from the Arctic—that air-conditioning was first introduced to the American public. The Balaban and Katz movie theater chain installed an air conditioner in one theater in 1917. A decade later, some estimate that two hundred movie theaters had installed air-conditioning.[112] By 1931, engineers observed, "No modern theaters are being designed today without giving consideration to complete air conditioning systems."[113] Movie theaters were one of the first places that comfort air-conditioning was deployed,

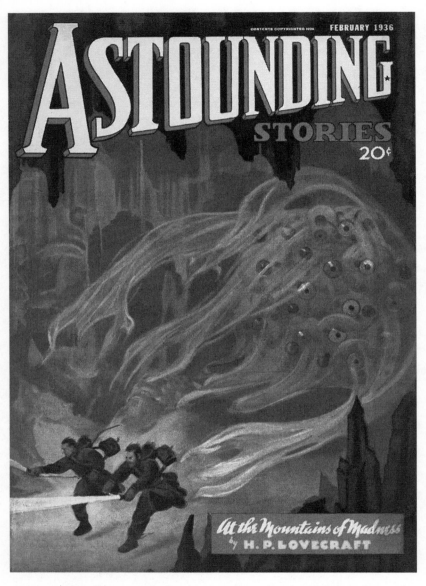

FIGURE 2.8. The illustration for "At the Mountains of Madness" depicts the racialized threat of the shoggoths as an atmosphere. From Lovecraft, "At the Mountains of Madness."

and as Willis Carrier, one of the inventors of modern air-conditioning, argued, theaters were "most helpful in making the public conscious of air-conditioning."[114] In *Air-Conditioning America*, Gail Cooper reflects, "More than any other market, the new motion picture theaters were the place where air-conditioning engineers and their customers happily created the illusion of a perfect artificial climate."[115] In turn, the air conditioner facilitated the expansion of moviegoing in the summer. The first technology of mass cooling was thus tethered to the cinema, the first technology of mass audiovisual media. It was during this period, when spectators were massed together as a visual audience in large-scale theaters and turned into a sonic audience through radio, that they also became a thermal audience, consuming cool air.

At the outset, however, for the cinema audience, the air conditioner was just one more medium that promised cooling, alongside cold films, atmospheric exploitations, fans, and ventilation systems. Unlike ice or ice cream, which cooled audiences from within (and which, theater managers acknowledged, could get messy and make people aware of their neighbors), the air conditioner is a convective medium, a means of manipulating subjects via the modification of atmospheres. For theater managers, owners, and engineers, it offered a particular kind of manipulation, one that—like the projector—was centralized and mechanical, and involved directing light and heat. It involved modulating a thermal source and broadcasting it to an environment.

During this time, theater managers and engineers considered thermal exposure and audiovisual exposures together as elemental and environmental practices. One engineer, writing in 1931, reflects back on the ties between cinema and air-conditioning: "When the production of motion pictures was beginning to assume the form of an industry in the early part of the present century, the art and practice of conditioning air was also beginning to emerge from the older, rule-of-thumb practices in heating and ventilating."[116] From the beginning, the author observes, the air-conditioning engineer and the motion picture engineer had worked together—and at the same time, helped to standardize and expand their industries. The art and practice of visual and thermal manipulation, engineers argued, depend on the separation, definition, and careful balancing of air, light, and heat between internal and external environments. Comparing the two, one remarks that the theater is an enclosure in which no light, sound, or "direct openings" were to be desired, while "the art of theater air conditioning" was dependent on bringing external air in, in part to "absorb the heat and the moisture that every person gives off and to reduce the odors that emanate from the body and the clothing of the patrons."[117] Like the arts of cinema, which required the creation of an enclosed world, with only specific lights and sounds allowed in, the air conditioner likewise negotiated transmissions in order to create a distinct thermal world and neutralize the ambient communications—heat, moisture, and smell—of others in the same theater.[118] In both cases, air is the medium of cinematic operation. Lecturing to the British Kinematograph Society in 1933, Thorne Baker reminds his audience: "Without air there would be no sound. . . . Not only is sound caused by a mechanical movement of air, but an air stream will convey *impressed* sounds in much the same way as a wireless carrier wave will convey an impressed message. . . . In both studio and theater, therefore, the problem of ventilation and the provision of air of a convenient temperature and humidity, must be coupled with that of noiseless *air*."[119] A Carrier Corporation engineer tells motion picture engineers explicitly, "We are considering air as a medium through which we may warm or cool the human body."[120] Long before media theorists began to describe the landscape, the body, and the elements as media for communications, air-conditioning engineers explicitly crafted air as a medium and cold as its content.

Although air-conditioning engineers and theater managers agreed that air was a medium, they were in conflict over what it was supposed to mediate. The air-conditioning engineers circulated the comfort zone chart to theater managers, offering a rubric according to which air communications should be designed and received. It was frequently published and discussed in theater operators' trade magazines. But, as in the deployment of the thermostat to the home, these standards—and the thermostatic subject they imagined—sat in direct conflict with cold exploitation and with vernacular practices that prioritized affective communication and the conscious manipulation of air as part of the cinematic experience.[121] As part of their cold exploitations, theater operators intensified temperature as part of the attraction itself, regularly altering the "ideal" setpoints to make the motion picture theater even cooler; at times even "differences of 20° to 30° [were] maintained and [were] advertised as a drawing attraction."[122] A Carrier engineer, frustrated with these manipulations, complained that air-conditioning had much more to offer than "mere cooling during three or four sultry months. Theater owners, even some of those who have wonderfully complete air-conditioning systems within their houses, have been prone to think of and term the equipment a refrigeration system. They play this feature up to the public in frosted letters nicely arranged about pictures of polar bears and icebergs. Fine!—psychologically, but not without its unfavorable reactions. Furthermore this type of advertising was used long before any theater had a system capable of cooling."[123] The ventilation engineers often argued that the air conditioner was *not* a refrigerator but a means to maintain "ideally comfortable and healthful conditions *every day in the year*"—a means of allowing everyone access to ideal thermal conditions.[124] In theaters, however, air-conditioning was never simply about making the environment livable and media viewable. Thermal manipulation was part of the consumable experience—a way of extending the marketing of coldness.

During this period, theater operators consciously used air-conditioning as a means of manipulating the "medium of air" to affect spectators' bodies. In motion picture industry periodicals, the use of thermal communication became a topic of extensive discussion, even when comfort rather than attraction was the guiding rubric. One author, writing on air-conditioning theaters in the *Motion Picture Herald* in 1937, observes, "One point that presents the greatest variety of opinions is the question of what are the ideal conditions." Even though, the article reports, most theater air conditioners were designed to create a 21°C/70°F environment with 50–55 percent relative humidity, "the limitations of the chart should be understood." It applied only to an average person living in the temperate zone of the United States, and "comfortable conditions are not standardized; people's habits, state of health, age, sex, clothing, and geographical location practically establish the conditions for comfort." The author then explains at length: "*Comfort* is merely a bodily sensation and is not absolute; it varies among individuals rather considerably. In applying air-conditioning no engineer really expects that all the patrons of a theatre are going to feel perfectly comfortable." It is here, however, that the author makes the direct correlation to cinema as an affective and communicative medium. "Does every patron of your theatre enjoy or like a particular picture to the same degree? Air-conditioning is just as intangible as entertainment; no single comfort standard can be laid down that will meet every need."[125] The audience is not conceptualized as just a set of viewers or listeners; it is a thermal audience, there to experience temperature, and audience members come in with a set of thermal desires and preferences.

As was true for the fans and ventilation systems that came before them, air conditioners could not simply be installed. They had to be engineered as media, especially through the production of the idea of thermal contrast. As one engineer argued in 1927, "The average person is conscious of comfort principally by contrast. That is why it has been easy to attract the public from the hot streets. . . . At times when outdoor conditions do not bring this contrast boldly to the attention of the patron the only comparison we have is to bring to his mind the fact that he *is* comfortable in comparison to some experience which he has had in a stuffy overheated audience."[126] The engineer articulated to theater operators what the Honeywell thermostat marketing study of the early 1950s would later discover: that it was necessary to call upon memories of discomfort, of heat, in order to make the cold legible and to sell temperature. Likewise, elaborate exploitations—with the same thermometers, icicles, and fake snow used to market coldsploitation films—had to be used to market the air-conditioned theater.

Even when they did utilize the comfort zone chart, theater operators often drew on their own understanding of cinematic mediation, thwarting the HVAC engineers' conceptions of the standard thermal subject. They used the chart flexibly, adapting it to different locations: theater charts were locally adapted to Miami, New Orleans, and Phoenix, among other cities.[127] Some operators varied the controls in concert with the ebbs and flows of the day, the particular audience, and the external environment. And still others argued that they were "not particular about how warm or cool you keep these people"; they were more concerned about modifying the temperature to prevent loafers: "Too comfortable in the lobby [and] they may not want to go inside."[128] In a series of articles "devoted to a better understanding of temperature control in theaters," Gordon H. Simmons reflects, "Your air-conditioning system for best operation must have 'tentacles' to take a feel for the outside weather and inside crowd." In turn, "feelings by the tentacles" must be translated into a proper thermal environment.[129] Railing against operators—whom Simmons perceived as misoperating systems, dropping them to temperatures that were too cold—the article argues that operators were giving air-conditioning and its engineers a bad reputation: "Those guys that make the insides of theaters so uncomfortable!"[130]

Although most widely deployed in cinemas, cooling technologies were also connected to the aerial transformations of radio—especially as both artificial cooling and radio moved into public spaces. For example, in the installation of air-conditioning on trains, the communication of cold was sometimes seen as an analogue to radio communication. The Chesapeake and Ohio Railroad equipped the lounge cars of "its finest train" with both radio receiving sets and air-conditioning, marketing both as aerial forms of manipulation. In a story about their advertising strategy, "Air-Conditioning on the Air," the C&O boasts that, recognizing the "vital part which radio plays in American life," they not only allow their passengers to tune into the broadcast but also use radio advertising to tell the story of a "genuinely air-conditioned train."[131] Perhaps not surprisingly, given these overlaps, some radio companies—seeing air-conditioning as an art of experience—also moved into the air-conditioning business. Philco Radio and Television Corporation in 1938 announced their entry into the air-conditioning field: "Like radio, air-conditioning is strictly a twentieth-century development—and like radio, air-conditioning will soon become a requisite for the average home."[132] The Cool-wave, offered by Philco and manufactured by the York Ice Machinery Company, was a single-room home unit intended to compete with the new air-conditioned theaters—to generate an affective space of thermal leisure in the home. "The men and woman who have enjoyed the comfort of an evening in an air-conditioned theatre are no longer content to return to swelter in hot bedrooms," the president of Philco claimed.[133]

The thermal manipulations of coldsploitation spanned sound, image, text, and air circulation. They amplified a sense of cinema's power of manipulation and its capacity to generate fantastic worlds. But the trend did not last very long. An initial cycle of coldsploitation films in the 1920s was followed by a revival in the early 1930s with *With Byrd at the South Pole* (1930), *Igloo* (1932), *90° South* (1933), *Eskimo* (1933), and *S.O.S. Iceberg* (1933), among others. From 1933 to 1935, Richard Byrd's second Antarctic expedition relayed radio messages from Little America, his base, to CBS, sponsored by General Foods. *The World Adventurers Club* radio show fictionalized polar expeditions in broadcasts such as "The Frozen North" (1932). But by the late 1930s, snow exploitation pictures had lost popularity. Exploitation and street stunts in their "broad and cacophonous forms" had disappeared.[134] And almost all cinemas had air conditioning. For many historians, this is a story of the triumph of comfort, of air-conditioning engineers who advocated a standard setpoint over theater managers who used air-conditioning in affective and illogical ways. For others, it is a story of cinema's seasonal expansion and the creation of summer moviegoing. While cold certainly did have an impact on cinema, its deployment was never simply about making the cinematic media viewable; steady temperature had never been a precondition for media engagement.

But even as they receded from popular media, cold communications did not disappear; they simply became an ambient sensory context layered into viewers', listeners', and media users' skin. It was through thermal media of the air—cinema, radio, and air-conditioning—that cold attractions were transmitted, sensed, and naturalized in bodies. In turn, coldsploitation proved the marketability of atmospheric communications—underscoring cinema's transformation of the public into a mass of spectators, colocated in a theater, and radio's work to create the airwaves as a form of communication. This period thus marks a transition to a new form of thermopower, one in which viewers' bodies were calibrated with the expectation that they would be able to consume such thermal attractions as part of mass media experiences.

Even after the coldsploitation cycle, when snow pictures and cooling systems were no longer a subject of concentrated advertising for cinema, this form of thermopower remained. It remained latent in the photography of Antarctic expeditions, in the televisual circulation of midcentury Arctic research, and the racist portrayals of Indigenous communities. It was elicited in the advertisements for the cold weather clothing and Arctic radio dramas of the 1940s and 1950s. Cold attractions surfaced again and again throughout the twentieth century in this mode: the creation of cold atmospheres through both representational and technical means, intended to draw in the bodies of media users, listeners, and viewers, to generate both pleasure and profits. While colonial travel often had a thermal rationale, coldsploitation linked thermal attractions to tourism. Today, the World Tourism Organization designates climate "an essential resource."[135] Cold attractions ultimately energize and facilitate cold weather tourism, often still defined by the consumption of thermal Otherness. As bell hooks so clearly explains, the "commodification of Otherness has been so successful because it is offered as a new delight, more intense, more satisfying than normal ways of doing and feeling."[136] These transformations all follow in the wake of the transition to convective mediation, in which air, wind, breezes, and environments are modulated to produce cold attractions.

FROM CONVECTIVE TO CONDUCTIVE MEDIA

In recent years, the atmospheric has been losing ground in its monopoly on thermal attractions. With the transition to adaptive, personalized cooling technologies (sometimes dubbed "personal comfort systems"), there has been a revival of conductive media—direct bodily cooling through touch—rather than convective cooling through atmospheric means. The ice vest, which in its modern form was constructed to enable military and industrial work in hot environments, is now a commercial product worn by athletes, cold therapy patients, and people looking to lose weight. Cooling neckbands and ice collars draw from a long traditional of using ice as a storage medium to deliver a blast of cold. Smart mattresses, smart mattress pads, and smart sheets allow people to control the temperature of their beds. Many of these new forms of digital, personal, and adaptive cooling work through direct contact. They put the body itself in touch with objects and use thermal interfaces to integrate bodily matter into their thermal circuits.

Over the course of the twentieth century, even with the turn to the atmosphere, the affective use of ice and cold water never disappeared. As the circulation of ice water and ice cream in 1920s coldsploitation media reveals, conductive thermal media were still enlisted as part of atmospheric cold attractions. But this more recent thermocultural transition capitalizes on a different medium of transmission—electricity rather than ice. Personal cooling systems often rely on thermoelectric technology, leveraging a process known even in the nineteenth century: that temperature differences can be converted to electricity (the Seebeck effect) and electricity can be used to produce temperature differentials (the Peltier effect). Much of the scientific and technical effort has historically focused on the former: converting thermal differences into electricity. Thermoelectric generators, which produce thermopower, produce electricity when there are different temperatures at either end of a device. From the nineteenth century onward, thermoelectric generators were used for a range of purposes: radio transmissions in rural areas in the Soviet Union, the generation of power on space missions, and the operation of cardiac pacemakers. At times, they were even used for consumer technologies (for example, in the late 1990s, Seiko developed the Seiko Thermic, a watch powered by body heat). Today, environmentalists identify immense potential in thermoelectric generation. As a scalable mode of energy production, it can be used to convert heat— whether the heat of industrial processes or the heat of human bodies—to electricity.[137]

On the other hand, the Peltier effect—the use of electricity to generate a temperature differential—was not taken up nearly so widely. In a review of this history, researchers show that it was not until the development of the semiconductor, an integral technology of computation, that a "thermoelectric breakthrough" occurred and the use of electricity to cool environments was seriously considered.[138] Even then, it only resulted in a few failed consumer products, such as the Westinghouse thermoelectric refrigerator in 1959. One place that the Peltier effect did become important, however, was in the development of optoelectronics, "where accurate and precise temperature conditioning was of primary importance."[139] Thermoelectric cooling, also known as "solid-state cooling" because it used no fluids or liquids, was useful precisely because it is not convective—and does not endanger electronics with liquids.

But even so, it was only with advances in nanotechnology and the development of a range of new thermoelectric materials that the Peltier effect could be harnessed for everyday consumer-based thermocultures—especially to develop a new set of cool attractions. The Embr Wave, a wristband with thermoelectric technology, is one such device. Like many other adaptive thermal technologies, from smart thermostats to participatory thermal sensing, the watch enables users to personalize their experience of temperature. Powered by electricity, it mobilizes the Peltier effect to cool or heat a spot on a user's wrist, providing them with "thermal relief." Rather than heating or cooling the environment around its users, the Embr Wave—like the ice collar—targets a specific spot on the body to convey cold sensations. The Embr Wave is explicitly a communicative device—as its marketing material even testifies, it doesn't change the core temperature of a body but instead "tricks" the user into "perceiving a slightly different temperature."[140] Like the smart thermostat, the device is marketed as an environmental and economic choice, one that also enhances comfort and productivity and allows users to regain control of their temperature *without* control over the environment. One user testimonial claims, "I can't control the thermostat in my office, but Embr Wave helps me stay comfortable and productive—no matter what."[141] The rationale for this technology, even more so than for the smart thermostat, is the acknowledgment of thermal difference, the fact that different bodies experience temperature in different ways and have differing levels of thermal access. Here, too, the device operates in service of thermal capitalism, promising maximized productivity.

The Embr Wave is just one of many new cooling devices that use thermoelectric technology. Sony's Reon Pocket, a "personal air conditioner," embeds a digital device into the back of a specially designed shirt (conditioning the body rather than the air around it). Thermoelectric coolers are stitched into car seats, mattresses, and clothing. As with adaptive thermal technologies, scientists have identified immense potential in thermoelectrics for addressing global warming. In a 2019 article in *Nature Communications*, researchers write that of the sustainable technologies developed to reduce the energy spent on cooling, "thermoelectric (TE) cooling is the most promising option as it provides pathway for personalized cooling."[142] This might seem like a return to the history of ice—a cooling mechanism that worked primarily and directly on the body itself. And in an even longer history, it hearkens back to the use of clothing itself as a technology to heat and cool bodies.

Interwoven with digital systems, controlled through mobile interfaces, and enmeshed in the language of adaptive thermal comfort, the Embr Wave and other digital conductive media evidence a new thermocultural trend of cool attractions, with a set of fantasies that diverge from the atmospheres of twentieth-century coldsploitation. Rather than immersing the body in an Other climate, these devices promise to isolate the body as a thermal medium from its environment, to make it less susceptible to atmospheric manipulations. In this new fantasy of the cold, digital systems that people operate are not heating up the earth—they are here to cool us down. Nor should our investment be in infrastructures or atmospheres—this is a thermoculture of the body itself, of pleasure that arises not from Bachelard's "gentle heat" of "man made warm by man" but from man cooled by machine.

New conductive technologies, including the Embr Wave, are also starting to feed back into thermal comfort research. Researchers have begun to conduct experiments using these thermal media, which in addition to correcting "non-neutral states" also provide the "opportunity of thermal alliesthesia."[143] This is not the first moment that thermal alliesthesia has become a site of inquiry. In the 1990s, when thermal comfort researchers began to critique thermal universalism, a second line of critique interrogated the stability of thermal meaning, especially of comfort. These researchers, in a classic poststructuralist move, pointed out that the scale of thermal comfort itself was rarely questioned. "Maybe we should be asking about preference directly," one thermal comfort researcher prompts as an alternative. "It is sometimes said that in summer, or in hot climates, people prefer to feel cooler than 'neutral.'"[144] Neutrality itself may not indicate comfort, the researcher observes, and discomfort can also be caused by unanticipated changes that might otherwise be comfortable. These researchers began to investigate the possibility of alliesthesia—a phenomenon described by Michel Cabanac in 1979—and explored whether thermal pleasure correlated with the internal state of the person perceiving temperature.[145] They cited Lisa Heschong's descriptions of thermal delight and suggested that thermal comfort might shift to become more expansive and include thermal pleasures. While in the years that followed, thermal multiculturalism gained a popular vernacular and was taken up as a rationale for a microcast model of individualized thermal reception, even with advances in research on alliesthesia, thermal neutrality remains the dominant paradigm.

While engineers still largely define the goal of thermal media as the achievement of nonstimulation, the history of the air conditioner, polar representations, and technologies of personal cooling demonstrate that temperature has been an attraction all along. While ice cubes, colonial travel, and the creation of spas, among other thermal practices, laid the groundwork for thermal attractions, from the late 1910s to the 1930s temperature became a mass media attraction—one that diverged from the personal devices and visual technologies of the nineteenth century. Air-conditioning thus doesn't emerge out of nowhere—it builds on, layers into, and intensifies an existing means of selling atmospheric coldness. And in turn, thermal capitalism does not simply tap into and satisfy a universal, biological, or environmental desire for particular temperatures: it creates them using the thermal media. As new technologies move away from air-conditioning and atmospheres and toward an environment in which people are cooled not by consuming ice but through contact with screens and materials, these new thermal media reveal the beginnings of a new ideology of coldness. Here, the human body is the subject of phase transitions, altered not simply through environmental controls but through the transformative work of digital systems.

3   SWEATBOX THE THERMAL VIOLENCE OF WEAPONIZED HEAT

In Saskatoon, Canada, police officers picked up a young person in the middle of the night. It was the end of November in 1990. The sky was blowing snow, and the temperature was almost −29°C/−20°F. Neil Stonechild was left by these officers on the outskirts of town and was found days later, frozen and missing one shoe. There was no hard evidence of murder, despite the strange marks on Stonechild's wrists and face, and the case was closed quickly. The investigating officer at the time simply speculated, "The kid went out, got drunk, went for a walk, and froze to death."[1] It would happen again and again. A decade later, another person would be picked up, driven far from any source of warmth, and left in the cold. But Darryl Night's story was the anomaly. Navigating to a power plant, to a taxi, and finally home, Night survived and spoke out about the incident, prompting an inquiry into these "starlight tours." As legal scholar and Chickasaw Nation member Sákéj Henderson describes it, these trips were a very old, racist, settler practice "to get rid of the Indian who was inebriated or mad."[2] Canada's white settlers were shocked to hear about these thermal violences, but for Indigenous people in the area, the fact that the environment could be weaponized was widespread cultural knowledge, stored in their bodies and evoked for some in the nighttime walks in the cold.

In the mid-1860s, a thirteen-year-old girl whose name is recorded only as "Miss B." was brought to an infirmary in Belfast, Ireland. Like so many other young women of the time, she was diagnosed with hysteria, the "woman's disorder," based on her "irritable habit," "indisposition," and occasional "delirium."[3] Over a century later, feminists would identify these "symptoms" as a "specifically feminine protolanguage, communicating through the body messages that cannot be verbalized."[4] Treatment generally included an array of violent tactics: sensory deprivation, seclusion, sexual assault, and sedation, and hydrotherapy (dousing people with extraordinarily hot or cold water). But for Miss B., the physicians decided to try an approach that extended the water cure. They sponged her with ice water and applied ice to her body, manipulating Miss. B as a thermodynamic object and using the extraordinary cold to attempt to subdue her "overstimulation." Reporting on their "success," one physician proclaimed, "The advantages derived from the employment of ice in any nervous disorders cannot be overrated."[5] Ice was used as a conductive medium to temper and cure disease, particularly sexual dysfunctions, curing some women of "the most violent nymphomania" and some men of "the incorrigible habit of masturbation."[6] For those defined as sexually deviant, expressivity was met with a violent cryopolitics. They would sense a dominant order through its thermal corrections.

At the United States' border with Mexico, immigrants to the country were detained in *las hieleras*—the iceboxes. There were no beds to sleep on. There was no space to lie down on the hard concrete. Some people tucked into a fetal position to stay warm. Others huddled together, but it was like "trying to sleep on ice."[7] There were no windows to allow the flow of air. The food was restricted and even the water was cold. They were stripped of their warm clothes and only sometimes given a Mylar blanket. They felt like they were freezing. Their lips grew chap and started to bleed. "My sister was shaking from the cold. Her lips were always blue," one girl recounts.[8] The officer told the children to stop crying or else they would "turn up the air conditioning and make it colder."[9] This cold has been the first experience of hundreds of thousands of people upon entry to the United States, as they were detained in cells that even the US Customs and Border Protection (CPB) admitted are not meant for long-term holding. Even though the CPB stated that the temperature stayed at 21°C/70°F, the many stories, reports, and class-action lawsuits revealed that air-conditioning has been used as an affective, convective medium, a means of sensory recalibration and physical harm.[10] "It's as though they wanted to drain every positive feeling out of us," one man recounts.[11] And the effects, the authorities hoped, would be contagious. When immigrants to the United States left las hieleras, they carried not only the memory of the cold but often respiratory infections and illnesses that had been incubated in that space. Their bodies and immune systems had been weakened. While concern flooded across the news media when a seven-year-old, Jakelin Caal Maquin, died after being detained, the illnesses of many others have gone unreported. These are just the latest instantiation of a larger program of "Prevention through Deterrence," in which border regimes use the environment to kill.[12]

In late November of 2016, the cold was weaponized once again. The water protectors had been encamped at Standing Rock for months. The weather was getting colder, and this made everything more difficult. On a night when it dipped into the twenties, North Dakota state police sprayed the protectors with a water cannon. Ice crystallized everywhere. Some people held up plastic tarps to protect themselves and the fires they relied on for warmth. Some bore the full brunt of the water cannon's pressure. Some were transported to hospitals and treated for hypothermia. Images of the spray shooting into the air, encasing human bodies, and forming icicles on barbed wire were circulated via social media, but to little effect. This tactic had been used many times before and would be used again. Just a few years prior, in Kyiv, Ukraine, police blasted activists with water cannons in freezing weather. One person died of pneumonia.[13] Over a century before that, the Homestead Steel Works in Pennsylvania was fortified with pressurized pipes to spray boiling water on its own striking workers.[14]

At Standing Rock, the snow began to fall. The North Dakota governor issued an executive order, "Emergency Evacuation to Safeguard against Harsh Winter Conditions." It states, "Winter conditions have the potential to endanger human life, especially when they are exposed to those conditions without proper shelter, dwellings or sanitation for prolonged periods of time."[15] Police violence weaponized the environment against its protectors. State authority used that violence as a rationale to remove these protectors, casting them as victims, not of police or capitalism or settler colonialism, but of the "harsh winter."

These are all instances of thermal violence.

Thermal violence is defined neither by a particular technology nor by simple exposure to extreme temperatures. It is a manipulation of a body's capacity to mediate heat. Thermal violence inhibits and pulls, constrains and extends, blocks and fills. It is a means of altering the body as a medium, either making it transmissive or forcing it to store excess heat. As a result, thermal violence is akin to other forms of torture by media, such as sound cannons that damage the hearing of protesters and strobe lights used in prisoner interrogation. And like the techniques of psychological operations, thermal violence is often described as a communicative and affective tool, intended to convey first and foremost an impression of temperature. These affective impressions—the messages of temperature—are characterized by the indeterminacy of heat effects, the ongoing difficulty of locating the precise effects of temperature on people's bodies despite an immense amount of research to this end. Just like other forms of communication, the exact reception and impact of thermal media escape scientific prediction.

As it manipulates the body's internal dynamics, thermal violence enables perpetrators to deflect blame from themselves to the environment. The police officers who left people in the freezing cold were not convicted of murder. The immigration officers who kept people in las hieleras were not held responsible for their deaths. Those who execute these techniques harness both the power and the excuse of climate. This environmental deferral is possible in part because of the indeterminacy of heat effects. The fact that different bodies thermosense and react to temperature differently opens a loophole for denial. Moreover, most thermal violence does not look like a burst of freezing-cold water, a structured conduit of thermal transmission. Thermal violence is often a "slow violence" that, as Rob Nixon describes, is "neither spectacular nor instantaneous" but accretes over time, often incrementally and out of sight.[16] Thermal violence's scarce traces and indeterminate effects make it difficult to address. Violence needs to be legible, visible, and documentable in order to make calls for accountability.

Many thermal violences, when described at all, appear to be "no-touch" techniques that allow perpetrators to harm their victims without direct contact. Even the water cannon works at a distance, weaponizing the environment and the ambient temperature to intensify thermal exposures. Perpetrators of thermal violence enlist ecologies, systems, and elements, often as convective media, in order to alter the capacity of bodies to emit heat and maintain thermal states, which often increases their vulnerability to other phenomena. They function by placing bodies in the "thermal margins," as Joanna Radin and Emma Kowal articulate: "The zones of precarity, ambiguity, and unexpected generativity that also reorganize ideas about what it means to be and remain alive."[17]

While the cases above are all instances of thermal violence enacted through the cold, this chapter focuses on an architecture for overheating: the sweatbox. I describe the long history of the sweatbox in the United States, from its use on southern plantations as part of slavery's infrastructure to its use in prison cells today. This is an explicit case in which communication via heat is an intentional activity—where there is a conscious manipulation of the elements, the environment, and architecture as media for thermal transmission in order to alter bodily activity. But the long history of the sweatbox also makes apparent how techniques of thermal violence, working on the body as a medium, reproduce and accentuate inequity not only because they are intentionally used to harm particular racial, ethnic, and religious groups but also because they differentially affect bodies according to social position.

Thermal violence is a systematic means of weaponizing environmental phenomena, deploying thermal affects to modify and regulate behavior, punish subjects, and exploit and intensify existing vulnerabilities. As it enables environmental deferral and relies on indeterminate effects, thermal violence has been an effective means of naturalizing racism, which, as Ruth Wilson Gilmore writes, is "the state-sanctioned or extralegal production and exploitation of group differentiated vulnerability to premature death."[18] While the standards of thermal comfort described in the first chapter were shaped by test experiments on white male subjects, racist, colonial, and patriarchal scientific studies of bodies in "extreme" climates have been used to naturalize and justify these forms of thermal violence.

By bringing together such disparate moments—of frozen fingertips and overheated bodies—the metallurgical approach of this book risks erasing the differences of racial, colonial, gendered, and sexual forms of oppression. Articulating a string of horrific events also risks perpetuating what Eve Tuck calls the "damage narrative": by focusing on harm and injury, these narratives risk creating a one-dimensional portrayal of people and communities as broken and depleted.[19] This chapter focuses on the sweatbox and connects it to a long history of racism intentionally and strategically in this context. I recount this history in order to recognize the harm and the specific mechanisms of thermocultural practices that are incredibly violent but not always treated as such. The perpetrators of these acts almost never held accountable or prosecuted. The effects of this violence on people's bodies remain largely invisible. And perhaps most significantly, in political and legal contexts, these cases are often personalized rather than indicative of systemic oppression. Thermal violence all too often remains in the background. As Tuck observes, this is the "paradox of damage: to refute it, we need to say it aloud."[20] Describing moments of thermal violence as part of a longer history and system of oppression exposes how supposedly neutral and universal thermal technologies perpetuate racialized, sexualized, and gendered forms of harm, and as I describe below, offers new approaches to addressing these forms of harm.

The history of sweatboxing related in this chapter illustrates how thermopower operates not only through normative practices and affective consumption but through the violent conditioning of one's capacities as a thermal being. Acts of thermal violence are less often perpetuated through the direct and intentional heating up or cooling of bodies than through the manipulation of movement and stillness, resources and proximities, architecture and labor. While previous chapters reveal how thermal subjects are created through the operation of thermal infrastructures and the commodification of thermal affects, this chapter draws attention to a wide range of ways that thermoceptive regimes are established, not simply through thermostats, air-conditioning, or even images and sound. People are heated through the provision of food, which fuels their metabolic activity. Their bodies are cooled by wind and water, which help to remove heat from the skin. They are warmed through activity, through participation in frenetic forms of sociality that can lead to overheating. The environment, inhabitation, and the entire realm of social practice is mobilized to condition one's body as a thermal sensor and receiver. In other words, bodies are calibrated as thermal media through a multitude of forms of social practice.

What the cases above and the pages that follow reveal is how thermopower exists in the ability to strip away people's capacity to manage and sustain their own body temperature. It exists in the diagnoses of bodies as incorrectly temperatured and in the forced recalibration that follows. It exists in the delivery of excessive thermal affects, even when the actual temperature isn't altered (thermal affects have been weaponized as often as they have been sold). It exists in the use of temperatures to justify the occupation of people's ancestral lands through genocidal and violent removal. And perhaps most importantly, it reveals that, while climate change will devastate environments, especially the environments of people already enduring violence within these structures of power, it will also expand the technological capacity and the available media for enacting thermal violence. In short, as the thermal contexts of existing architectures and inhabitations are altered, the environment is being transformed into a massive sweatbox for ambient, environmental, and intimate means of shaping thermal subjects. The latter half of this chapter, which focuses on the prison cell, describes one place where this is already occurring.

I recount these forms of harm here in order to ground a new politics of temperature that can counter such practices. While the first chapter questioned the project of a universal thermal subject and the second chapter questioned a universal sense of comfort, this chapter questions whether the current formulation of exposure, the dominant paradigm for describing thermal harm (which is based in racialized studies of the body), is sufficient to contest thermal violence. The conception of thermal exposure (the contact between a discrete body and external environment) often models thermal exchange on a classical form of communication, where there is direct channel, a sender, and a receiver. As I describe in the chapter's final pages, despite the advocacy that has been successful within this paradigm, the common sense of thermal exposure is limited in a number of ways: it fails to account for the durational effects of thermal violence, and it often ignores actual bodies, assuming a subject without race, gender, or sex. The final section of this chapter suggests that we must broaden our understanding of exposure. A body is not a photograph; its exposures accrete over time, calibrating sensation. I suggest that a politics of thermal autonomy can make these invisible affects and coercions visible and better account for the vulnerabilities of bodies and the racial regimes of thermal violence.

SWEATBOX

The box is only slightly larger than a person's body. It is a vertical coffin. There is not room to sit or rest. One has to stand up. There are holes for air circulation, perhaps one inch by four inches each, and water might be received once a day. This architecture appears to separate the prisoner from the guard, a protoform of no-touch torture, but the sweatbox is designed to penetrate bodies and manipulate their temperature. It uses the materiality of wood to intensify the sun's rays, restricts water that might let prisoners perspire, forces them to stand (and exert energy to balance), and limits the capacity for respiration that releases heat from their internal organs. It is a thermal medium that uses air, water, wood, metal, and the body itself to limit one's ability to emit heat into the world. The sweatbox would typically be placed in a cold, damp place during the winter. While in the summer the design of the sweatbox restricts transmissions from the body, consolidating heat within the skin, in the cold and wet the same set of materials transforms the body into a conductor. Rainwater could be used to release the heat in the body. Immobilization and withholding food are tactics to restrict metabolism and, in turn, the body's ability to generate heat. The sweatbox, in other words, is not simply a cell of confinement that happens to be placed in a hot or cold environment. It is an architecture that reorganizes the very capacities of one's body to mediate heat, a thermal violence of the most intimate kind.

The sweatbox was not invented on the plantations of the American South—it was used by colonizers to torture Native Americans as well as to punish Tory offenders during the American War for Independence—but the form described here crystallized in its development by southern enslavers.[21] Despite the scarcity of testimony by enslaved peoples, the narratives that do exist suggest that in the antebellum period enslavers used the sweatbox to punish significant transgressions, and that both men and women were imprisoned in it. Although less often represented than other forms of physical torture, the sweatbox was deeply imbricated in the South's racialized landscape, a part of what Darla Thompson identifies as the "infrastructure of slavery."[22]

In an interview with the Works Progress Administration in 1936, Prince Smith, who was formerly enslaved, recounts that the sweatbox was so small that a person would have to be squeezed in. In the summer it would be put in the hot sun and in the winter in the coldest and dampest location.[23] Another formerly enslaved person recalls that in the sweatbox they would be fed cornbread without salt and just plain water. They might be kept in the box for months, and when released, they would scarcely be able to walk.[24] The details of these peoples' accounts reveal that the sweatbox was not simply a physical shell around the human body—a cage constructed by enslavers—but a set of techniques of withholding and restriction. As a result, the sweatbox could materialize even when there wasn't an architecture built specifically for the purpose of subjugation. Enslavers reappropriated existing architectures toward this end—a wooden barrel or a "screw box" used to press cotton.[25] Even beyond these architectures, enslaved Africans would often be housed in "poorly constructed slave shacks that admitted winter cold and summer heat," an extension of sweatbox conditions.[26] Although it might have appeared to observers as a discrete architecture, the sweatbox was a deeply embedded set of white cultural practices that used temperature to help maintain enslavers' "absolute monopoly of violence."[27]

As a result, compared to other practices of physical torture, the sweatbox seemed to some a lesser form of punishment. Janie Scott, recalling others' experience on an Alabama plantation, observes that the overseer was not mean because there was a "sweat box to punish unruly slaves in place of whipping them."[28] This comparison foreshadows later arguments that the sweatbox was a more humane alternative to other forms of physical punishment. But such statements were infrequently made by those who had been subjected to the sweatbox. Although articulated in a radically different context, Kang Chol-hwan's description of imprisonment in North Korea in the 1970s and 1980s offers details of the brutality possible with the sweatbox. Chol-hwan recounts that the sweatbox was "one of the harshest punishments imaginable" and that it broke even the sturdiest of constitutions. It was possible to survive it, but the cost was often crippling and the aftereffects were almost always permanent. It was simply grisly: the privation of food, close confinement, crouching on one's knees, hands on thighs, unable to move. The prisoner's rear end pressed into their heels so unrelentingly that their buttocks turn solid black with bruising. Chol-hwan observes that hardly anyone exited the sweatbox on their own two feet.[29] Although a person is not physically touched in the sweatbox, their body is permeated and could be debilitated to the point of scarcely being able to walk. In chattel slavery, whips and brands left marks that testified to the violence that was inflicted, and these marks were crucial as instruments of management. In contrast, those outside the sweatbox often viewed it as a lesser evil, in part because of the invisibility and the indeterminacy of heat effects.

Although the sweatbox was instrumental on the plantation and shaped by its use by enslavers, it also emerged in the key institutional contexts of the nineteenth century, including schools, prisons, and ships. Following the 1877 riots at the State Reform School for Boys in Westborough, Massachusetts, details were revealed about the sweatbox torture of boys.[30] Likewise, mid-nineteenth-century reports called attention to the use of the sweatbox in prisons, including New York's Sing Sing.[31] The sweatbox was also used as a tool of naval discipline.[32] A surgeon in 1858 on the USS *Dale* described a case of a young person placed in a sweatbox for six hours without food or water. The surgeon detailed the architectural dimensions of the box (approximately six feet seven inches by seventeen and a half inches by fifteen inches) and the occurrences before and during confinement: "The man had already been engaged with the ship's company in exercising sails," the surgeon writes. "The day was one of the warmest we have had," and "the cells are close by the ship's galley, which, of course, much increases the temperature."[33] These circumstantial elements are necessary to "understand fully the severity of the punishment," the surgeon argued. What this case points to is that the sweatbox was not necessarily harmful by itself but rather when deployed at a particular time, for a particular person, and in a particular environment. It is a contextually harmful architecture of violence. For the young person on the ship, the sweatbox extended from the walls of the cell to the lack of food and water, to the abnormal heat of the day, to the activity engaged in previously. All these things intensified the heat already contained in the body.

In the postbellum period, the sweatbox continued as a means of punishment and coercion across the American South, especially in the prisons and on the chain gangs that formed the new plantations. Even though the sweatbox could be used on white prisoners, by and large it was a means of racialized violence. It was often used selectively, to punish some prisoners and not others. In many places, Black prisoners were more likely to be thrown into the sweatbox than white ones.[34] One journalist comments in the late 1920s that the "sweat box was devised, not for the whites of Florida, but for Black Florida citizens."[35] Indeed, as Alex Lichtenstein describes, corporeal punishment, including "confinement in a 'sweat-box' under the southern sun . . . was meted out for the most insignificant transgressions, particularly to African-Americans."[36] This could cause swelling in the legs that led to hospitalization.[37] Death from sunstroke and from suffocation in the sweatbox was not uncommon.

In the 1910s and 1920s, some reformers of the prison system endorsed the sweatbox in place of "direct" corporeal punishment, even as this practice continued to disproportionately harm and kill Black people.[38] In Texas, after the 1910 abolishment of convict leasing, state legislators hired a reformer, Ben E. Cabell, as the chief prison administrator—Cabell favored the sweatbox as a "humane alternative to the strap."[39] Only a few years later, eight Black prisoners suffocated to death in one Texas sweatbox—a higher number of deaths, Robert Perkinson observes, than in any single incident under convict leasing.[40] In the early 1920s, following the post-flogging death of white prisoner Martin Talbert, Florida governor Cary Hardee outlawed the flogging of prisoners in the state.[41] As in Texas, this prompted a turn—or a return—to the sweatbox as a means of punishment, defined as "a cell with solid walls . . . '3 feet wide, 6 feet 6 inches long and 7 feet from the floor to the grating over the top' and 'so constructed that it can be divided across in two equal parts, and a convict may be confined in one half of the space in the day time, but shall have the full space in the night time.'"[42] This architecture was used, as it had been in years prior, to murder Black prisoners, including Henry Ridley, who was killed in a sweatbox in Florida in 1927. The *Chicago Daily Tribune* ran only a brief report: "A coroner's jury decided that Ridley came to his death by natural causes, the exact nature of which was not known."[43]

It was not until the 1930s that the sweatbox garnered national attention, thanks to its representation by white authors and the confinement of white prisoners. In January 1931, a sweatbox appeared in Robert Elliot Burns's *I Am a Fugitive from a Georgia Chain Gang!* The following year, Jim Tully's *Laughter in Hell*, the story of an Irish American railroad worker sentenced to a chain gang, also featured a sweatbox. And only a few months later, the death of the white prisoner Arthur Maillefert, nicknamed "Jersey" for Maillefert's home state, was followed by a sensational, nationally reported trial. Restrained within the sweatbox by a chain around the neck and wooden stocks around the feet, Maillefert was strangled to death only shortly after being placed inside. The "chain gang hanging" occured at the Sunbeam Prison Camp in Florida and became a touchstone for North-South tensions. Following Maillefert's death, northern news outlets launched a campaign to abolish the chain gang that was flooded with antisouthern sentiment.[44] In a landmark, and anomalous, decision, one prison official was convicted of manslaughter and sentenced to twenty years in prison.

As the use of the sweatbox became a national political issue, catalyzed by white authors and white prisoners, the sweatbox appeared in the chain gang cycle in Hollywood, including the low-budget *Hell's Highway* (September 1932), *I Am a Fugitive from a Chain Gang!* (November 1932), and *Laughter in Hell* (January 1933).[45] Subsequent to the release of *Hell's Highway* and the publicity surrounding Maillefert's death, a group of white convicts went on strike at a road camp following an attempt to place one prisoner in a sweatbox.[46] When Black people died in the sweatbox, however, few paid attention. John Spivak's fictional narrative about Black prisoners, which ends with the protagonist being thrown in a sweatbox, was overshadowed by the Maillefert case. There was very little press coverage of cases involving Black prisoners, such as the sweatbox murder of Lewis Gordon on August 12, 1941. In the decision of the trial following Gordon's death, "the judges said the jury was authorized in finding that the 'proximate cause' of Gordon's death was his confinement with 21 other convicts in a small wooden building 'without adequate air or water.' According to the evidence, Jacobson [the prison guard] placed the convicts in the building and kept them there from 1 P.M. to 8 P.M. while the temperature outside was 105 degrees. The warden was indicted on a charge of murder, but the jury convicted him of involuntary manslaughter."[47] As in cases like Henry Ridley's, the deaths of these people were often attributed to an accident or "natural causes," their murder obscured and the guards' agency deferred to the environment.

While this history is broadly indicative of racism in the United States, the sweatbox emerged as a set of thermocultural practices in this context to become a particular tactic of racial violence, a practice that was deemed more appropriate for Black people than white people. What set the sweatbox apart from other techniques of racist violence during this period, such as lynching, is its invisibility and indeterminacy. It is foundational to what Darius Rejali calls the "American style of stealth torture," developed by the 1920s and defined by a preference for technologies that left no physical mark.[48] Rejali and others track the transmutation of sweating and police interrogation through Stalin's conveyor system, the German interrogation of British airmen during World War II, and the American torture of prisoners at Guantanamo. In parallel, the use of the sweatbox also continued throughout the twentieth century in the United States, and especially in the American South, as a means of bringing about the premature death of Black people. That history—the way that the sweatbox pioneers a form of racialized violence that can remain invisible and escape direct contestation—is much less frequently documented.

The sweatbox, and sweating, as a form of interrogation was used during the American Civil War to gather military information, often by locating a hot stove next to a cell.[49] The first widely reported instance of sweating as an interrogation technique was of John Stapler, a Black man who was arrested in 1868 and kept in a sweatbox for thirty hours in order to obtain a confession. After the event, Stapler spoke at public events, becoming well known for the sweatbox incident. From the late nineteenth into the early twentieth century, the sweatbox became an important technology for police interrogations, and its use escalated following World War I.[50] Sweating a suspect, a practice broadly used by the police, was a highly racialized tactic. In the "sweat box case" of 1902, the Mississippi Supreme Court threw out the conviction of Edward Ammons, whose confession was deemed involuntary after being kept in a sweatbox during the heat of summer. In a discussion of the case, Christopher Waldrep describes the police work of softening up "black suspects for questioning in a five-by-six foot box, or 'apartment,' covered in blankets to deprive inmates of even a stray ray of light or a breath of fresh air."[51] Prisoners were allowed to communicate only with interrogators.

These forms of thermal discipline were legitimated by the dominant "knowledges," both scientific and popular, about Black people during the nineteenth and early twentieth centuries. During this period, theories of climatic determinism linking climate and race and debates about the dynamics of acclimatization circulated widely.[52] These originated in Western thermal theory but took on a particular nature in relation to chattel slavery. Some variations of climate determinism suggested that Black people's bodies took on too much heat and thus became volatile. Counter to this, another dominant white paradigm, described in critical studies of medical racism, endowed Black people with a resistance to heat. Plantation medicine was used to legitimate slave labor: enslavers and the physicians who worked for them spoke about Black peoples' ability to work in the hot sun, even though, as Harriet Washington shows, "planters' own records bear the best evidence that blacks' 'immunity' to heat prostration was merely a convenient myth."[53] The use of the sweatbox, in its many manifestations, in the US South is aligned with and legitimated by both of these racist beliefs: on the one hand, Black people were seen as the most appropriate and receptive for thermal conditioning; on the other hand, they were understood to be immune to thermal transmissions.

This history of the sweatbox illustrates several key elements of thermal violence in the United States. First, thermal violence emerged as a systematic and racialized tool for disciplining Black people, first by enslavers and later across the institutional contexts of the nineteenth and twentieth centuries, especially in policing and the prison system. Across these contexts, even though it appears less intrusive than direct corporeal punishment, the sweatbox functioned as a means of intimately affecting bodies while allowing authorities to escape accountability: deaths caused by the sweatbox were frequently recorded as deaths by "natural causes," not murder—the lack of volition associated with thermal exchange enabled this environmental deferral. Given the use of the sweatbox on plantations and the ongoing racism in the United States, deniability was not essential for this technique to emerge. But, as I describe in the next section, these practices pioneered a mode of weaponizing the environment that became much more widespread and ever more accessible in an era of climate change.

Second, the power of the sweatbox depended not simply on the exposure of a person to thermal phenomena but rather on the manipulation of their ability to retain, mediate, or release heat. This was done not simply through wood and metal or air-conditioning but through the manipulation of movement, position, and food. The key operation of the sweatbox was either to concentrate heat, transforming one's body into a storage medium for heat, or to quickly transmit cold through one's body. This shifted one's capacity for life, altering a person's breathing, metabolism, and awareness.

Understanding the sweatbox as a racist thermocultural practice rather than a discrete architecture makes clear that even after the three-foot-by-three-foot box disappeared, sweatboxing persists as part of what Saidiya Hartman describes as "the afterlife of slavery."[54]

PRISON

Outside Jacksonville, Florida, not far from the former location of the Sunbeam Prison Camp, where Arthur Maillefert died in 1932, the legacy of the plantation sweatbox lives on in the Union Correctional Institution. Formerly called the Raiford Prison and the Florida State Prison, it was established in 1913 and was one of the last to abolish convict leasing, in 1923 (many prisoners at that point were simply transferred to the prison's farm, a state-run plantation).[55] Today, the prison is part of a complex stretching for miles, encompassing several more-recent units and acres of cattle ranching. The Union Correctional Institution is a prison-agricultural complex in which the labor of incarcerated people powers a food system; the beef cattle emit nitrogen, consume massive amounts of water, and contribute to global warming; and the prison's forestry projects, also worked by prisoners, decrease the number of trees, which absorb carbon. In turn, the changing climate expands the human capacities for thermal violence against these same prisoners.

Even after the national antisweatbox advocacy of the 1930s, the six-foot-by-three-foot sweatbox at Raiford continued to be used as a means of punishment. In the 1940s, another wave of sweatbox opposition swept the area, but it remained relatively local, focusing specifically on the prison. This followed the death of another white prisoner, a military veteran. Through the 1950s, prisoners were reportedly kept in the sweatbox for over a week, given only water and a half pound of cornbread (figure 3.1).[56] After the official use of sweatboxes stopped in 1958, guards adopted other means of sweatboxing. Marvin Johnson, who arrived at Raiford in the 1960s, recalls, "Inmates could be put in 'the box,' a small outdoor cell that guards closed off during the day, so the temperatures would rise to 110 degrees, and opened at night, so the mosquitoes could feed."[57] At Raiford, as at other prisons across the country, the "official" sweatbox architecture disappeared only to be replaced by the prison cell itself.

This is nowhere more true at Union Correctional than on the death row block. As David Martin, a prisoner in this unit, succinctly puts it, "If it's Summer, it's extremely hot on the wing, while during Winter it's extremely

FIGURE 3.1. Sweatbox at the Florida State Prison (now Union Correctional Institution), Raiford, Florida. From Duane Perkins, State Library and Archives of Florida, Reference Collection RC05716, 1957.

cold."[58] Death row is not simply an architecture of isolation separating people from their surroundings. It is an architecture of environmental intensification. In the late 1990s, to sit on Union Correctional's death row block during the summer was to be baked. This was especially true during the 1998 heat wave, during which almost 1,500 maximum temperature records were tied or broken across Florida.[59] On death row, there was no air-conditioning. There were no circulating fans. The inmates were prohibited from having personal fans. They did what they could to keep cool, using various materials (often the cardboard backing of legal pads) to deflect air from vents onto themselves. Prison officials then banned the use of these "air deflectors" and stopped running the ventilation system. The restriction and strict definition of air flow into and out of the cell, a key feature of sweatboxing, limited the inmates' ability to modulate their own body heat. They were confined to their cells nearly twenty-four hours a day, allowed out only a few times a week for showers and exercise.[60] The cattle down the road on the prison farm received water and were free to move to cooler areas. But death row prisoners were held still as heat accumulated in their bodies.

The prison guards did not build the cells. They did not turn up the heat. Nonetheless, they engaged in a set of practices of sweatboxing: limiting movement, limiting water, and regulating food in ways that amplified the heat stored in prisoners' bodies. As was true for the nineteenth-century sweatbox, thermal violence is enacted not simply through confinement but through a set of thermal techniques and practices. Climate change does not overheat the prisoners; it merely expands the ease with which the guards can do so.

In 2000, the death row inmates filed a class action suit against the prison officials, alleging that high temperatures in their cells during the summer amounted to cruel and unusual punishment. After a denial and an appeal, the court concluded, "According to accepted engineering standards for institutional residential settings, the temperatures and ventilation on the . . . unit during the summer months are almost always consistent with reasonable levels of comfort and slight discomfort which are to be expected in a residential setting in Florida in a building that is not air-conditioned."[61] Here, thermal comfort standards, which defined a uniform level of comfort for a non-air-conditioned building, were drawn for the prison as a legitimation for its thermal practices. In this ruling, the multitude of ways that temperature and heat are manipulated (through adjusting water, food, movement, air circulation, the capacity to change position, shade and sun, and so on) were ignored or considered irrelevant. As in earlier moments, when enslavers used sweatboxes instead of traditional corporeal punishments or when the prison cell replaced the wooden box at Raiford, the limited understanding of thermal violence as enacted by a built architecture or exposure to extreme temperature obscured the harm.

During this period and in the years that followed, the American Civil Liberties Union's National Prison Project mounted several cases against prisons, several of which involved some form of sweatboxing. There were a few small successes. In 2002, after a hunger strike by inmates on death row at the Mississippi State Penitentiary—which, like Union Correctional, was set up as a prison plantation in the South (it was established in 1901 on the site of an existing plantation) and as a means of upholding white supremacy—the ACLU uncovered numerous instances of sweatboxing.[62] The court ordered "immediate remedies" for these practices.[63] In 2003 in Baltimore, a court ordered the declaration of a "heat emergency" when temperatures in a women's prison exceeded 32°C/90°F for more than four hours and mandated the provision of air-conditioned bed space for prisoners at risk for heat-related illnesses.[64] The following year the US Court of Appeals for the Seventh Circuit ruled that officials must cool the prison cells at a facility in Boscobel, Wisconsin.[65] The judge decided: "Defendants constructed a facility in which inmates are subjected to temperatures that can pose a serious risk to their well-being. . . . If air conditioning is the only means of avoiding that risk, that is a function of defendants' decision to build the facility as they did. Leaving inmates vulnerable to serious health consequences or death is not a reasonable alternative."[66] In this anomalous ruling, the prison officials were held responsible for the thermal effects of the buildings they operate.

Despite these successes, antisweatboxing advocacy continues to face challenges specific to thermal violence. The temperature of prison cells and the effects of heat on the bodies of incarcerated people are rarely officially archived. During trials, medical experts visit and testify to prison conditions. Inmates aren't allowed to do their own thermal readings. But temperatures vary and rarely leave a physical trace, even in the case of death. There are of course exceptional moments when thermal violence is perceptible and publicized. Take for example, the first recorded use of air-conditioning in a sweatbox, during the civil rights protests in the Mississippi State Penitentiary in 1961, when inmates were soaked with a fire hose and subjected to air-conditioning for three days.[67] Or consider the 2012 murder of Darren Rainey, which occurred when guards at the Dade Correctional Institution in Florida locked Rainey in a shower, turned up the heat, and inflicted massive burns.[68] Such cases are exceptional in that they are documented and more widely circulated, whereas most thermal violence remains hidden.

The indeterminate, contextual, and largely invisible nature of thermal violence is what makes this tactic so powerful in the current moment—not only because of climate change but also because the media landscape makes *some* cases of carceral violence more visible to people outside. While pictures can capture images of cuts and bruises, thermal violence remains particularly illegible. Moreover, the acknowledgment of the thermoceptive body is used to legitimate thermal violence and to deny its impacts. In one CIA report, a source argued that the use of cold is not an interrogation technique "per se" and commented, "Cold is hard to define. He asked . . . 'How cold is cold? How cold is life threatening?'"[69] Even though there is an extensive amount of research and numerous guidelines defining how cold is too cold—much of which has been generated by the government itself—for the CIA, the contingency of the body is ultimately used to state that thermal effects vary widely depending on particular circumstances.[70] And this is backed up by the science of the thermoregulatory system, which observes the capacity of the human body "to survive exposure to a remarkably broad range of environmental thermal stressors . . . all while maintaining the nearly constant core temperature necessary for health and well-being."[71] This is the loophole that enables officers to rely on their own judgment and then deny they have any responsibility. All these features of thermal communication thus present challenges for those who advocate for change.

While there is a massive scientific apparatus that tracks global climate change, the influence of heat and cold on the bodies of those who will suffer the most remains understudied. Thermal violence, as it intensifies heat in the body and shapes a person's capacity for life, brings one closer to death in many ways that escape scientific analysis. As a result, critical temperature studies must look for trace indications of heat and ephemeral sensory experiences in the margins of existing accounts of social life and in the archives. One place where thermal violence is well documented is in prison writings, which make obvious the many ways that heat and cold are weaponized and how they differentially affect bodies.[72] Ramon Peguero, describing the suicide of a friend, writes: "I know he was stressed out. I was stressed out. We had to endure six consecutive days of near one hundred degree weather and the humidity raised the heat index making living in the cell unbearable. I reason, he just couldn't take it anymore. . . . Guys around here are dropping like flies. In the same week Bill hung-up another guy followed suit. Sadly, Bill was the third of four suicides since June 2010."[73] How many people buried in the prison cemeteries, whose deaths were recorded as suicides or "natural causes," were killed by thermal violence? A Texas prisoner reports, "During the summer time, it gets so hot in these red-brick ovens they call penitentiaries, men suffer heat-related deaths every year."[74] People in prisons today, just like the people who were thrown into sweatboxes on the plantations or in the chain gangs, know that the environment can be used against them. Thermoception is not simply present oriented; it is anticipatory. Some people hold with them an anticipation of the seasonal changes that can be harnessed to bring about premature death.

Many people in prison identify these thermal practices as part of white supremacy. Muti-Ajamu Osagboro writes:

> The attacks against your being while in prison are multifaceted in general, but in the hole those attacks are accelerated. The cells are brutally cold . . . constantly! The heat, or lack of it, is controlled by staff. My hands and feet are always like ice cubes, numb and with very little or no feeling. . . . "We're freezing. The air conditioner is blowing (full blast)."—Ronald Yandell, Hunger Strike leader at Pelican Bay State Prison. . . . Princeton's Dr. Cornel West, had it absolutely right when he said, "We're talking about something that is somatic. It's at the level of body and it's sonic, at the level of sound. We don't even like the sound of your name. You see that's White supremacy at a deep level."[75]

Like sound, Osagboro points out, temperature—coldness and heat—work at the level of the body, somatic attacks against one's being, making one's hands like ice cubes, withdrawing feeling from one's feet. Across the prison system, temperature control is a means of enacting white supremacy at a deep level, an intimate means of racial violence.

This description echoes a multitude of stories from others. In the 1978 essay "Women in Prison: How We Are," Assata Shakur begins, "We sit in the bull pen. We are all black. All restless. And we are all freezing. When we ask, the matron tells us that the heating system cannot be adjusted."[76] Analyzing this moment of thermal violence, Stephen Dillon connects it to a long past, writing that "affect continually forces the past to open directly onto the present. Frozen skin speaks in a way that words cannot. In prison, shivering black flesh weighted with chains looked like slavery."[77] The prison sweatbox, and the plantation sweatbox out of which it emerges, both hearken back to the thermal violence of the slave ship. Olaudah Equiano recounts of the ship: "The closeness of the place, and the heat of the climate, added to the number in the ship, which was so crowded that each had scarcely room to turn himself, almost suffocated us. This produced copious perspirations, so that the air soon became unfit for respiration, from a variety of loathsome smells, and brought on a sickness among the slaves, of which many died."[78] The slave ship becomes a sweatbox, producing perspirations and thermal emissions so thick that people's breathing becomes difficult and their bodies are made more vulnerable to all of the emissions in which they were immersed.

The effects of temperature are not simply physical or physiological. Heat and cold are filtered through thermoceptive regimes and highly localized activities, from the food people ingest to the movement they partake in on any given day. A body, as Joy Parr describes, is a "synthesizing instrument that defies the categorical and linear descriptions of language and science. . . . They are not only being *conditioned* by circumstances, they are also enduring reservoirs of past practice."[79] For this reason, heat effects appear indeterminate. But for those persons held in prison cells, thermal violence elicits a "sensorial attunement," as Marina Peterson describes, in which heat links "temporal moments materially and metaphorically."[80] Even as thermoregulatory experts visit the prisons to gather evidence for legal cases, they fail to grasp the ways that architectures open many of these incarcerated people not simply to extreme heat or cold but to a history of enslavement and thermal conditioning. This is an intimate form of violence intended to bring subjects into line, not via changes in ideology or verbal communication, but through the direct manipulation of bodily matters and affective states.

THERMAL AUTONOMY

In the prison strike of 2018, incarcerated people from across the United States and around the world advanced a set of demands, the first of which was for "immediate improvements to the conditions of prisons and prison policies that recognize the humanity of imprisoned men and women."[81] While neither temperature nor sweatboxes were specified in the list, the heat waves of that summer may have been in the organizers' minds. Reformers and abolitionists, like the ACLU, focus on defining temperature limits and mitigating practices, such as distributing water and circulating air, to take in the case of extreme temperatures. Defining limits to exposure is a crucial step in achieving practical relief for those who are thermally assaulted. A definition of thermal rights could not only help to alter prison conditions but also mitigate the effects of climate change on vulnerable populations more broadly. Take, for example, Lee Phillips's article "In Defense of Air-

Conditioning," which argues for a right to air-conditioning, even with its energy costs: people must have "free or cheap, reliable access to the thermal conditions optimal for human metabolism (air temperatures of between 18 degrees C and 24 degrees C, according to the World Health Organization). Neither too hot nor too cold."[82] Acceptable temperature norms could form a basis for institutions to engage in technological strategies, such as the extension of air-conditioning, to shield bodies from fluctuating temperatures. Such standards could be deployed across cultural, architectural, and institutional locations where sweatboxing and other forms of thermal violence are in operation. Take, for example, the regular overheating of students in Philadelphia schools, which lack air-conditioning: they slump in their seats until the heat causes an early dismissal.[83]

One aim of retelling this history of sweatbox operations has been to substantiate such claims. Here I describe how thermal violence has been present not only in the deployment of water cannons or in interrogation techniques but also in the everyday practices in fields and in prisons. It is an intimate, affective, and racist form of violence that penetrates deep within the body under attack, altering its capacities for mediation and for life. But this long history of the sweatbox also reveals some limits to the existing paradigm of exposure.

First, exposure is often circumscribed to the moment of violence, both temporally and spatially: harm seems to occur in the moment a beam hits a body and end when that body leaves a cell. In this model, thermal transmissions appear to operate via a sender-receiver model of communication. However, as I've described here, the power of thermal violence is as strong in its aftermath as in the moment of exposure. Bodily capacities can be invisibly shaped or manipulated in ways that cause ripple effects and conditioning for one's future. Affects can be drug up from the past. As Christina Sharpe writes: "Slavery was not singular; it was, rather, a singularity—a weather event or phenomenon likely to occur around a particular time, or date, or set of circumstances. Emancipation did not make free Black life free; it continues to hold us in that singularity. . . . [A]nti-blackness is pervasive as climate. . . . [I]t is not the specifics of any one event or set of events that are endlessly repeatable and repeated, but the totality of the environments in which we struggle."[84] The thermal violence of this past can become a form of thermal trauma that repeats, not as a blast or a shot, but as a set of ecological constraints and pulls and repetitions. Anti-Blackness is both "pervasive as climate" and circulating in the climate itself—in the heat, cold, and warmth that people encounter.

The existing paradigm of exposure also fails to grapple with thermal effects caused even within set temperature limits by practices of sweatboxing, including forced movement or stillness and the regulation of vital substances, which affect the body's mediating potential for heat. Most violent thermal effects do not occur through the actual or intentional alteration of climate. The existing paradigm of exposure also remains based on a conception of humans as distinct from their environment, and in many cases, the only possible framing is to see the institution as a means of protecting its prisoners from an external climate. This thinking mirrors many histories of technology, which tell readers it is through thermal management that civilization develops in the face of a harsh and threatening environment. As is clear at Union Correctional, most forms of thermal violence are historically shaped, culturally conditioned, and structural. It is not the sun's heat that's responsible for death but a re-engineering of elemental capacities for transmission, which in turn limit the degree to which bodies are able to emit heat. Lastly, the political and logistical challenges of extending thermal media, especially air-conditioning, in a moment of changing climate could position environmental advocates against prisoners and others who are made susceptible to heat through thermocultural practices.[85]

Yet when one moves beyond traditional conceptions of exposure, as many personal testimonies and scientific studies acknowledge, it is clear that thermal violence is best preempted by thermal autonomy: the ability to regulate and mediate one's positionality within the thermal world. Take, for example, Susi Vassallo's description in a report on risks of heat-related illness in the Mississippi State Penitentiary: "An individual free to respond to the stress created by a hot environment would normally take steps to cool his body. If no air conditioning were available, he would at least respond by seeking a cooler location, blocking out radiant heat from the sun by positioning himself in the shade or screening himself from the sun, maximizing evaporation by wetting his body and clothes with water and using fans to create cross ventilation, and moving away from physical structures which absorb and radiate heat."[86] The capacity to survive is tied to the fact that all bodies are thermal beings, continually emitting, regulating, and being affected by heat. We do not stop at our skin; our capacity to live, breathe, exist extends in and through the environments around us. As Stacy Alaimo has argued, human corporeality is transcorporeality, our substance enmeshed with the more-than-human world.[87] A feminist politics of exposure, as Alaimo articulates, requires a declaration "that humans are not outside the planet looking in, not floating above the phenomenon of climate change, but instead, that we are always materially interconnected to planetary processes as they emerge in particular places."[88] Documenting exposure involves tracking the "entanglement of . . . emissions with our bodies," Rahul Mukherjee points out.[89] In other words, this is an exposure modeled not on bodily contact with a separate thermal world but on thermal entanglement. From this perspective, granting thermal autonomy requires more than a change in cell or temperature: it requires a change in people's ability to determine the conditions of their own vitality. It requires an understanding of sweatboxing not as an isolated set of events but as a systemic and pervasive form that is part of a long history of exporting environmental harm to Black and brown communities.[90]

A politics of thermal autonomy makes clear that simply installing air conditioners in prisons is insufficient to counter thermal violence, just as the elimination of the "official" sweatbox architecture failed in this project. In both of these cases, although reformers advocated for improvements, each reform ultimately reinscribed a regime that enabled sweatboxing by other means. A politics of thermal autonomy thus also reveals that incarceration is not a neutral context in which thermal violence may or may not be enacted but a system that expands and multiplies the possibilities for thermal harm. Climate change is not a perpetrator of thermal violence but a phenomenon that increases the human capacity to weaponize architectures and environments. While a politics of exposure has traditionally been used to justify increased reliance on institutions and their technologies and architectures, thermal autonomy requires freedom from them. It does not require separation from "outside" temperatures but begins with the acknowledgement that people are inevitably entangled with the thermal environment.

In this way, thermal autonomy is the antithesis of the new temperature-control devices described in the previous two chapters. Even as these digital thermal media promise autonomy, personalization, and freedom from standards, they link users into a vast network of information exchange in which thermal data is leveraged for profit and bodies are increasingly subject to manipulation by technology and utility companies. Moreover, the technological calibration of personal temperature—whether via digital devices or the distribution of personal fans, as one Florida organization advocates—can do little to counter forms of thermal violence enacted through the vast array of thermocultural practices. Thermal personalization calibrates a person to expect control and to adapt one's own body through preference. In direct contrast to this, thermal autonomy ensures that people have the capacity to control the factors that shape their ability to mediate heat in the first place: movement, stillness, metabolism, positionality, and inhabitation. One cannot address thermal violence in an era of climate change by either eliminating sweatboxes or offering technologies of temperature control; one can address it only by changing the conditions in which climate can be weaponized.

PART II

4   HEAT RAY THE THERMAL CIRCUITS OF RADIANT MEDIA

In 2010, the United States military deployed a new technology to Afghanistan: the Active Denial System. Nicknamed the "heat ray," the system directs a millimeter wave beam at human subjects up to a half mile away. Electromagnetic radiation penetrates the human body only 1/64 inch, but it produces an extraordinary sensation of being burned—of being scalded by hot water or set on fire. The military fired thousands of test shots on volunteers' bodies and reported that while the system left no mark or burns, "the instinctive repel response [was] universal."[1] In the military's attempt to demonstrate the system's immaterial nature, the heat ray became one of the most-studied nonlethal weapons in the history of the Department of Defense, its precise effects documented by independent review boards, human-effects researchers, and technology specialists. These groups concluded that it was one of the least physically intrusive crowd control and security technologies and that, within established limits, it enacted no physiological harm on its targets—merely a sensation of harm, a sensation of heat. It was, some argued, the "Holy Grail of crowd control."[2] The military staged events to garner support for the heat ray, zapping reporters and demonstrating the absence of burns, scars, or debilitation. After it was recalled from Afghanistan, the system was redubbed the Assault Intervention Device and installed in Los Angeles's Pitchess Detention Center in order to weaponize heat domestically against incarcerated people.[3] There was public opposition, the American Civil Liberties Union contested its use, and the heat ray was ultimately decommissioned.

The Active Denial System encountered resistance in part because it evoked a deeply ingrained imagination of the heat ray—a thermal weapon of bodily disintegration. "It is still a matter of wonder how the Martians are able to slay men so swiftly and silently," H. G. Wells writes in *The War of the Worlds* in 1898. "However it is done, it is certain that a beam of heat is the essence of the matter. Heat, and invisible, instead of visible, light."[4] While Wells's fictional heat ray gun causes its target to burst into flames, the Active Denial System leaves no perceptible trace. As an instrument of thermal violence, the Active Denial System modulates behavior in a way that makes a victim's action look intentional, deferring accountability away from its operators. Even when transmitted directly from military installations, heat effects remain indeterminate, prompting speculation and interpretation.

I begin this chapter with the Active Denial System because even though it has much in common with the technologies described in the preceding chapters—invisibly orienting subjects, generating thermal affects, and enacting thermal violence—it materially diverges from the convective and atmospheric approaches of much twentieth-century thermal media. It is a radiant thermal medium. As I describe in the following pages, radiant thermal media generate, intensify, or channel electromagnetic waves to produce thermal effects. They can also directly leverage the infrared part of the spectrum for communication. While convective and conductive media transform the environment into a conduit for thermal communication (altering temperature through air circulation, water, interfaces, screens, minerals, architectures, and social practices), radiant thermal media organize materials to reflect, refract, or block spectral transmissions. Radiant thermal media are often supported by what Rahul Mukherjee calls "radiant infrastructures," large-scale infrastructural systems that directly emit radiation, and notably, they can transmit heat over a distance without heating the intervening substances.[5] Diving into the history of early twentieth-century experiments with radiant thermal media, this chapter exposes how thermopower operates not only through physical heating and cooling of elemental surrounds but also through the modulation of spectral waveforms.

Like the previous three chapters, this chapter documents the inseparability of thermopower from thermal media. Because radiant thermal media involve spectral negotiations, they are intimately connected to other forms of spectrum management: the manipulation of light waves, radio waves, and television signals. As a result, radiant-heat management and transmission systems are closely linked to existing wave-transformation systems: how we engage, imagine, and organize heat fundamentally reflects the landscape of other communications media. In turn, the politics and potential of the media spectrum inflect the possibilities for thermal organization.

In the early twentieth century, experiments with three parts of the electromagnetic spectrum—light, radio, and infrared—were intertwined and imagined as part of the development of new communications systems. Although infrared radiation had been sensed long before it was "discovered" in 1800, following the development of the light-based photophone in the late nineteenth century, infrared was imagined as a means of communication. Infrared circuits were later adopted as a means of television transmission, including in the early television systems designed by John Logie Baird. However, while radio waves became a substrate of mass media in the 1920s and 1930s, infrared waves were not broadly adopted as a means of media transmission. The early media history of channeling infrared is largely a story of failure, of attempts that were ultimately superseded by wave communications in other parts of the electromagnetic spectrum. As the infrared spectrum was populated with the promiscuous emissions of so many human and nonhuman heat producers, it was much more widely engaged during this period as a medium of sensing rather than a means of signal exchange.

At the same time that scientists and engineers were experimenting with infrared, radio manufacturers were experimenting with radiant heat. With the expansion of shortwave radio in the early twentieth century, people discovered that shortwave transmitters didn't simply send voices and signals—they also sent heat. Alongside experiments with infrared telegraphy and telephony, this catalyzed experiments with "radio fever": the use of shortwaves to heat bodies in medical practices, to cook food, and to transmit power. The experiments also generated new ideas about radiant heating in which subjects were targeted by heat rays from afar, rather than warmed or cooled by their immediate environment. These systems formed the foundation for the Active Denial System today, even if they used a different part of the electromagnetic spectrum.

The bulk of this chapter weaves through these technical histories and their corresponding cultural imaginations. After an overview of the specificity of radiant thermal communication, it, like the chapters before, focuses on the early decades of the twentieth century—especially the years following World War I. My aim here is not to offer a technical history of early infrared transmission—many books have already done so. Rather, it is to convey a sensibility, during this period, of what heat was and what it could become: a communications circuit. Like the Active Denial System, in this moment thermal communication was imagined as a direct channel by which wave-forms and, as a result, bodily activity could be modulated. Contrary to the operation of stoves, air conditioners, fans, and sweatboxes, all of which work largely via environmental modification and through the amplification of existing temperatures, in these experiments heat was imagined as a vector, a directional force that moved in time and space and was itself intended to produce directional movement. These systems channeled radiance into thermal circuits, transmitting heat signals and heat effects in a classic mode of communication, often to send messages directly and point to point.

These early twentieth-century experiments illuminated the affective and visceral potential of the infrared spectrum and composed a foundation for many subsequent infrared media technologies. But it was not until the mid-twentieth century, with the development of silicon media, plastic insulation, and lasers, that infrared signals were finally used in circuits of mass mediation. In the 1970s and 1980s, infrared waves came to constitute the beams of remote controls. Heat rays began to be used as "ink" to inscribe ubiquitous thermal paper receipts. And in the 1990s, consumer infrared linked computer laptops and personal digital assistants before being replaced by Wi-Fi and Bluetooth (two technologies that don't require line-of-sight connections). During this period, however, infrared still remained marginal as a mass transmissions medium.

In the twenty-first century, infrared signals became *the* substance of global digital media. The final sections of this chapter focus on the latest iteration of infrared communications: the radiant signals channeled down fiber optic cables. Inspired by the "optics" of these systems, those who write about and represent the cable network that carries most digital media traffic have often focused on "light" as the medium of transmission. But digital content—from social media posts to streaming video—is not encoded in visible light but in infrared waves. Massive amounts of data are transduced into heat rays and channeled across oceans and continents on hair-thin pieces of glass. Today's digital network is an infrared network.

A history of the heat ray, from the infrared telegraph to the infrared internet, shows how thermopower operates as a form of spectral manipulation.

As Helga Tawil-Souri points out, the spectrum is the "relational backbone to the devices and networks we build"—this is true for infrared as much as for the wavelengths used by radio, television, and Wi-Fi.[6] This history also reveals how the development of radio and television has generated new ways to produce heat effects, from the Active Denial System to the kitchen microwave. Through these experiments and infrastructures, the history of infrared is innately intertwined with the development of traditional communications technologies. Thermopower operates both through air conditioners and thermostats and through radio, television, and the internet.

As the infrared spectrum expands as a medium for signal exchange, the capacity to control infrared radiation shapes conditions for the global distribution of media. This in turn dramatically expands capacities for sensing and manipulating temperatures. Focusing on just one example of this, the closing section describes how infrared-based digital communications networks are being transformed into the most expansive thermal-sensing network on earth. New technologies enable infrared waves on fiber-optic cables to register temperature, with the potential to turn the global internet's infrastructure into a massive thermometer. While infrared radiation has long been entangled with telegraphy, radio, and television, climate change is now thermally entangled with the base substrate of the global internet. How people send infrared signals through the internet—and their dependence on digital platforms—materially expands the possibilities for sensing the warming of the planet as a whole.

RADIANCE

We are all radiant; we *radiate*. All bodies emit some kind of electromagnetic radiation as long as their temperature is greater than zero degrees Kelvin. The wavelengths at which bodies emit radiation depends on temperature. At higher temperatures, the particles inside a body are moving around with greater speed, and as result, the wavelengths of radiated emissions are much shorter. Exploding stars can emit X-rays because they are so hot. Colder bodies have less particle motion and thus emit longer wavelengths. The giant molecular clouds where stars are formed are some of the coldest places in the universe, with an internal temperature of 7–20 Kelvin. Scientists glean information about these clouds from the electromagnetic radiation they emit—in this case, long-wavelength radio waves. On either end of the spectrum, very hot and very cold phenomena across the universe can be measured by sensing their electromagnetic waves. For this reason, the investigation of the interaction between electromagnetic radiation and matter, called spectroscopy, has been critical to astronomy, weather reporting, and many other scientific practices. These electromagnetic emissions are also called thermal radiation and are described colloquially as heat.

While extraordinarily hot and extraordinarily cold phenomena in the universe emit radiation across the electromagnetic spectrum, almost everything on earth—from people's bodies to the clothes that they wear to the ground that they walk upon—emits infrared radiation, which is below the range of visible light, with just slightly longer wavelengths. Very hot things (a fire, an industrial smelter, the sun) emit radiation in the visible spectrum. This is why, once matter reaches a certain temperature (past a few hundred degrees Celsius), heat becomes visible. Whether conceptualized as waves (as in NASA's image, figure 4.1) or as a stream of photons, infrared emissions are continuous and ever present. They are, however, divided into four bands. The near-infrared spectrum, immediately proximate to the range of visible light, includes the shortest set of wavelengths, running from 0.7 μm to 3 μm (700–3,000 nm). The wavelengths in this band, band I, help to transmit communications between remote controls, for example, and are not perceived by people as heat. Wavelengths in the mid-infrared spectrum, called middle-wave infrared (MWIR) or band II, are longer, running from 3 μm to 8 μm (3,000–8,000 nm). Wavelengths in long-wave infrared (LWIR), or band III, also described as thermal infrared, run from 8 μm to 15 μm (8,000–15,000 nm). Human bodies tend to emit thermal radiation in this portion of the spectrum. Far-infrared waves are longer, running from 15 μm to 1000 μm (15,000–1,000,000 nm).

All beings are enmeshed in a field of radiant exchange—especially infrared exchanges. However, they are not equally receptive to or penetrated by these ambient emissions. The radiation emitted by any given entity is not a direct reflection of temperature alone. It also depends on the entity's surface composition. In calculating how emissive something is, physicists refer to what they call a "black body," an ideal figure that perfectly absorbs and emits all electromagnetic radiation. The "black body" is a perfect medium: it takes in radiant waves and relays them without change, interruption, or decay. In reality, surfaces and composition matter. They affect the movement of thermal radiation. This feature is called "emissivity," and each kind of surface is given a number that describes how much its composition decreases its capacity to transmit radiation. While the black body, as a perfect emitter and a frictionless medium, has an emissivity of 1, all real matter has a value less than this.

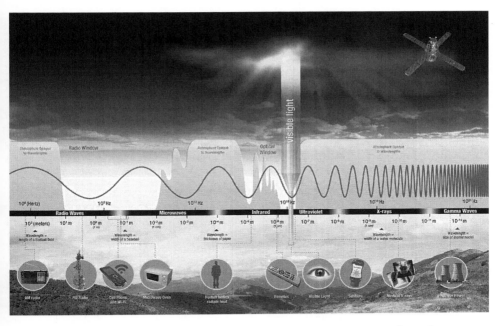

FIGURE 4.1. The electromagnetic spectrum as explained by NASA. From "Introduction to the Electromagnetic Spectrum," NASA, accessed April 30, 2019, https://science.nasa.gov/ems/01_intro.

Substances that come close to this ideal are entities like ice, which has an emissivity of 0.97, and rubber, which has an emissivity of 0.95. This means, roughly speaking, that ice will absorb and emit roughly 97 percent of the thermal radiation in its environment. It both takes on the radiations of other entities and releases its own radiation. Ice is a good medium—it is conductive. On the other hand, materials such as aluminum foil (which has an emissivity of 0.03) and polished silver (with an emissivity of 0.02) reflect much ambient thermal radiation back to their surroundings and are imperfect emitters themselves. Rather than conducting, they insulate, preventing the flow of infrared waves. Complicating this is the fact that emissivity itself varies with temperature: a given material's ability to mediate is shaped by how hot or cold it is.

Just as cameras manipulate the wavelengths of visible radiation and musical instruments manipulate sound waves, radiant thermal media manipulate this milieu of infrared radiation through management of materials, emissivity, and position. On the one hand, radiant emissions—and their radiant energy—have long been harnessed as forms of heating. Both the sun and fires produce visible and infrared emissions—these are sources of radiant heat. Indoor radiant-heating technologies warm bodies in their homes. Even the design of clothing manipulates spectral emissions, including both the emissions of one's body and the emissions of the sun. Most clothing, especially when made from natural textiles, has an emissivity of 0.7–0.95, and this gives it a relatively high capacity to transmit thermal radiation.[7] Some researchers argue that dipping fabric in a silver nanowire solution, and thereby drawing on silver's extraordinarily low emissivity, would enable a form of "personal thermal management."[8] The infrared radiation emitted by a person wearing clothing dipped in the silver nanowire solution would be reflected back to them. Like the Nest thermostat and the Embr Wave wristband, high-tech clothing is a frontier in thermal personalization. An "anti-drone" art project, Stealth Wear, incorporates silver-plated fabric into garments, so that they will contain infrared radiation emitted from the body and reflect external radiation. The media of fabric and silver are used to block thermal visibility.

The fields of architecture and textile design, among others, have generated extensive research on emissivity and thermal management. These concepts, and the manipulation of infrared radiation, have been central to the last century of building and garment construction. And yet, while media and communications scholars have studied the spectrum—the use of radio waves, television waves, 5G, and even X-rays—there has been little work on infrared waves as a substrate of communication, whether in fashion, architecture, or networked media.[9] Understanding heat and cold requires, in addition to analyses of conductive and convective media, an attention to the specificity of radiant media. While conductive media facilitate the transportation of heat directly through contact, and convective media use substances like water or air for thermal transfer, radiant media communicate heat through the manipulation of spectral activity and the dynamics of conductivity and emissivity. When water cannons are used as a form of thermal violence, water—which has a high emissivity—coats a human body, speeding up its radiant emissions to the outside world and thus accelerating cooling. Experiments in building insulation in the 1970s, described in chapter 5, use low-emissivity materials to contain thermal radiation—harnessing the bodies of the building's occupants into the heating infrastructure. These examples reveal that even when heat is not directly generated and the speed of molecules is not manipulated, thermopower organizes materials to reflect, refract, and block spectral transmissions. Through such modes of organization, heat waves—

ambient emissions in which all entities are enmeshed—can become heat rays, thermal circuits that are directional and intentional forms of transmission. Because heat occurs in a spectral form, like radio waves, it can be encoded with information. And since electromagnetic waves can produce heat effects, heat rays have an enormous affective capacity.

CHANNELING INFRARED

Unlike the waves of many other parts of the spectrum, infrared radiation is perceptible. Standing over a fire or a hot stove, people might feel its thermal emissions on their faces and hands. Working outside, people can sense the sun's thermal emissions. Taking cover in the shade, people allow the canopy to absorb the sun's radiation instead. Thermal media, whether windows or wool sweaters, have long been used to modulate radiant heat. In 1800, when William Herschel first identified the presence of infrared waves, this was not a moment of discovery so much as a new way to describe something most people already viscerally knew.

In this experiment, as would be true for many subsequent engagements with infrared, Herschel was initially studying visible light. In an attempt to determine the temperature of different-colored lights, Herschel moved a thermometer past the edge of visible spectrum. Here, just beyond the color red, the highest temperature was actually invisible. Herschel named this radiation "calorific rays." But Herschel's contribution was not simply to name these rays. Conducting experiments with radiation from a fire, a candle, and a hot poker, Herschel found that the rays obeyed the same laws as visible light. They could be reflected and refracted. In this moment, the substance of radiant heat was linked to the substance of light.

These calorific rays, to be later renamed "infrared radiation," were the first form of electromagnetic radiation beyond visible light to be "discovered," and this opened the spectrum more broadly to scientific scrutiny. Still, it was more than half a century before James Clerk Maxwell developed a theory of electromagnetic radiation, and Heinrich Hertz did not demonstrate the existence of electromagnetic waves until 1884. Their work, in turn, generated a multitude of experiments with wave communications in the late nineteenth and early twentieth century, from the manipulation of visible light as a communications signal to the encoding of radio waves to the production of television signals. Alongside these many spectral experiments, infrared rays were also imagined as a medium of communication that could be used to relay sounds and images.

One of the first media to leverage light waves as a form of communication was developed by Alexander Graham Bell and Charles Sumner Tainter in 1880. The photophone, like the telephone and the phonograph, was designed to transmit sound. It did so, not via a wire or a cylinder, but "by the agency of a beam of light."[10] The photophone apparatus was fairly simple. On the transmitting end, a person would speak into a tube. The sound waves of their voice would hit a flexible mirror, via which the "vibrations of a sound . . . throw into vibration a beam of light."[11] The vibrations in this beam of light—natural or artificial—were then registered by a photovoltaic selenium cell, which turned the light into electricity. Via a reversal of this process, electricity was transduced back into sound and registered by a listener. Bell described this process of speaking through the air via light: "I have heard articulate speech produced by sunlight! I have heard a ray of the sun laugh and cough and sing!"[12]

The photophone and subsequent optical experiments depended not simply on the manipulation of radiation but on the discovery and investigation of the spectral sensitivity of matter, especially the light sensitivity of a particular element: selenium. Selenium is sensitive to the visible spectrum of light, and photovoltaic selenium cells can register transmissions from the ultraviolet to the infrared, with maximum sensitivity in the visual range: 350 nm to 750 nm.[13] The selenium cell's peak spectral response occurs at 556 nm, almost exactly the spectral sensitivity of a human eye. When exposed to light, selenium conducts electrons—it converts radiant waves to electric energy. Selenium is not only photovoltaic but also photoconductive, which means that as light increases, its conductivity increases (when illuminated, its conductivity could increase more than a thousandfold).[14] These properties would later make selenium a crucial part of solar technology, integral to the first practical solar cells and to the earliest light-exposure meters.

In the nineteenth century, the discovery of the "action of light on selenium" occurred during experiments with a communications system—as Willoughby Smith tested its usefulness for submarine communications cables.[15] This discovery catalyzed experiments with the photophone, which in turn generated more research on the spectral sensitivity of matter. Bell and Tainter concluded in their photophone experiments that there was an entire class of substances, from paper to gold, that would emit sounds correlating to vibrations of light.[16] In other words, these investigations were not simply into the spectrum itself but into matter's responsiveness to it.

The photophone is a radiant medium, one in which light waves are channeled and transduced into signals as part of a communicative apparatus. The invention enabled radiophony via light well before the discovery of radio waves and the practical deployment of wireless telegraphy. Combining the photophone with a spectroscope, a technology that senses fluctuations in radiation, Bell also developed the spectrophone, which allowed its users to listen to infrared waves. Bell writes, "The ear cannot for one moment compete with the eye in the examination of the visible part of the spectrum; but in the invisible part beyond the red, where the eye is useless, the ear is invaluable."[17] The photophone not only established a basis for hearing infrared for the first time but for early imaginations of heat rays. H. G. Wells was a reader and contributor to the journal *Nature*, in which Bell published "Selenium and the Photophone." The description of the heat ray in *The War of the Worlds* (1897) mirrors the operation of the photophone: "Many think that in some way they are able to generate an intense heat in a chamber of practically absolute non-conductivity. This intense heat they project in a parallel beam against any object they choose, by means of a polished parabolic mirror of unknown composition, much as the parabolic mirror of a lighthouse projects a beam of light."[18]

The photophone was debated in scientific journals and discussed in newspapers. It was exhibited publicly at events like the International Electrical Exhibition in Philadelphia in 1884 and the Chicago World's Fair in 1893. But while wireless telegraphy spread, the photophone was not taken up as a mass medium. World War I, with its need for developments in secret signaling practices, returned attention to the photophone, and photophone technology was a focus of researchers in England, the United States, and Germany. For most of 1916, at University College London, Alexander Oliver Rankine conducted experiments with light-wave communication for the Royal Navy's Board of Invention and Research. Like Bell, Rankine used selenium as a primary detector, but he found it limited. Rankine argues, "If a suitable substitute free from [selenium's] inertia-like effect could be found it would very soon displace selenium cells and the similar devices at present in use."[19]

During the war, a new detector was developed that enabled invisible infrared signaling. In the United States, Theodore Willard Case created a thallium sulfide cell, which was much more sensitive to infrared radiation and therefore more suitable "for covert infrared communication."[20] The cells were manufactured at Case Research Laboratory in Auburn, New York.

Filters manufactured by Eastman Kodak blocked visible light but allowed infrared rays to come through. These materials were incorporated into a series of prototype heat-ray transmitters for military use. But, as Case described after the war, "It was realized . . . that there would be no demand for such apparatus or its further development except for war purposes, consequently it was decided to stop all further work along this line."[21]

In the early 1920s, as atmospheric cooling was being pioneered in cinema and people's homes were permeated by radio, the infrared telegraph briefly circulated as an object of popular attention. The September 1920 issue of *Popular Science Monthly* reported on a "heat ray telegraph" developed by French soldiers.[22] It described the technology as a thermal photophone, one that, like Case's apparatus, used infrared radiation to wirelessly transmit signals. *Popular Mechanics* also featured extensive discussion of the "infrared-ray telegraph"—with a thermoelectric cell made of tellurium rather than selenium (figure 4.2).[23] But even as Rankine and Case began to publish their research, the photophone, an instrument with a "chequered career," still largely remained a "novelty."[24] Aside from some experiments with the device for transmitting sound in cinema and experiments by amateur radio enthusiasts, the photophone—and direct communication by light and infrared waves—failed to become a mass medium.

As the experiments in heat ray telegraphy and telephony dwindled, another inventor began to channel the spectrum for transmission, not to send sounds, but to transmit moving images. In the winter of 1926, John Logie Baird, often credited with the invention of television, offered a demonstration of a new technology. In a laboratory in Soho, London, Baird pointed the Televisor at the head of a ventriloquist's doll and transmitted its image to a screen for a live audience. The Televisor worked using the same principles as the photophone and the heat ray telegraph (Baird was inspired by H. G. Wells's novels). The person (or doll) being televised was illuminated with intense floodlights and subjected to their glare and heat. Bouncing off of these televisual subjects, the light moved through a rotating disk and came into contact with a series of photoelectric selenium cells—a "television eye."[25] The cell generated an electric current in proportion to the light to which it was exposed. This current was then converted to "ordinary" wireless signals or sent through a wire and, via a reversal of the process of capture, produced an image.

Listening to the early subjects of television, Baird heard complaints of sitters "dazzled and blinded by the brilliant illumination."[26] Indeed, Baird's very first televised subject ran away from the bright lights. It was difficult

Complete Infrared-Ray Receiving Set: A Three-Stage Amplifier at the Left, and the Detector, Potentiometer, and Interrupter at the Right

At the Left Is the Small Portable Set, Consisting of Searchlight on Tripod, Storage Battery, Thermoelectric Detector, Interrupter and Head Telephone. In the Center Is the Horn-and-Ear Tube Receiver. At the Left Is a Close-Up View of the Three-Stage Vacuum-Tube Amplifier

FIGURE 4.2. The infra-red ray telegraph, 1920. From "Infra-red Ray Telegraph Has Interesting Equipment," *Popular Mechanics*, May 1920, 649.

to reduce the light's intensity and still retain the image's quality. As a result, Baird decided to try "invisible rays."[27] Using ebonite to filter out the visible light from the illumination, Baird found that infrared rays, like the light rays, could be used to transmit images: "At first I used electric fires to produce these infra-red rays which are practically heat rays. I could not get a result and added more fires until Wally was practically roasted alive. Then I put in a dummy's head and added more fires until the head went up in flames. I decided to try another tack and used the shorter infra-red rays."[28] After replacing blinding lights with roasting infrared rays, Baird eventually tried placing hot plates next to the faces of the television subjects. By the end of 1926, Baird offered another demonstration to Britain's Royal Institution: a televised image of a person sitting in darkness. What Baird called "Noctovision," and what is today called "active infrared imaging," emerged in these experiments with television. Infrared was harnessed for image transmission precisely because its rays would not interfere with vision.[29]

The use of infrared promised a vision of the invisible, and heat's revelatory potential was sexualized. One of the earliest Noctovision images recounted by Baird captured a sexual assault. On a radio broadcast in 1931, Baird described the "interesting little episode": while watching the face of one of the female staff members on the screen, Baird witnessed a second face enter the image and kiss the woman. Baird asked the engineer about it afterward, and he replied "that the temptation of the dark room and the good-looking young lady had been too much for him."[30] In 1933, during tests with an infrared technique, the BBC "'stripped' cotton dresses off a line of dancing girls."[31]

The use of infrared rays was not ideal for traditional television—and not only because of these potential revelations. The photoelectric cells skewed color: "Red appeared as white while blue did not appear at all."[32] Beyond this, it was incredibly hot. The intense floodlights and the infrared systems were eventually replaced by a scanning light. As a result, even though the Noctovisor—and infrared television—was groundbreaking, one historian argues, "It really represented no advance for television."[33] Other televisual technologies took center stage, and ultimately the use of infrared as a transmissions medium was abandoned.

Nonetheless, infrared waves remained a site of research and experimentation. Military researchers continued to investigate infrared communication. The British Admiralty undertook experiments on infrared's potential for fog penetration through 1934, as it was "a subject of considerable importance to safety at sea as well as signaling at sea during poor visibility."[34] The possibilities for infrared waves multiplied. Physicians had long used infrared radiation informally, but they began to think more specifically about infrared rays as part of various treatments. In 1930, William Annandale Troup released a book dedicated to the topic: *Therapeutic Uses of Infra-Red Rays*.[35] And as light therapy moved from the hospital and the clinic to the home in the 1920s and 1930s—as Tania Anne Woloshyn documents in *Soaking Up the Rays*—the infrared lamp emerged on the mass market.[36] Infrared beams were also incorporated into sensing mechanisms. In their investigation of the famous medium Rudi Schneider, Dr. Eugene Osty and son Marcel connected a photoelectric cell to an infrared beam, which was used to guard an object in a dark séance room. When the infrared beam was disrupted, a photograph was taken—but this early infrared trigger system provided no evidence of deceit.[37]

Just as Herschel's discovery of infrared wavelengths fundamentally linked infrared emissions to light emissions, these early media experiments often imagined that infrared rays could be used like visible light but would offer a means of communication unhindered by visibility. These inventors saw infrared communication as a potential form of telephony, telegraphy, and television. Many assumed that infrared rays—heat rays—could function like sound waves. But Bell's, Baird's, and Rankine's experiments—even as they were foundational for military research, integral to medical practice, and set the stage for later infrared sensors and triggers—were not built out as commercial technologies or networks. Nonetheless, their work offered both a powerful imagination of heat rays as a medium of communication and a set of technical advances that would set the stage for digital networks almost a century later.

RADIO FEVER AND BROADCAST HEAT

As scientists, engineers, and militaries experimented with sending messages via infrared rays, a second kind of heat ray was also being developed—but in relation to radio waves rather than in relation to visible light. In the early 1920s, radio expanded so rapidly and was taken up with such enthusiasm that many described it as a kind of fever. "If any parent wants to know how to keep Johnny amused in the evening," one journalist suggested, "just have him contract the radio fever, which has become quite as prevalent as measles in the vicinity of New York, and involves no doctor's bills."[38] Long before "cool" provided the cultural sensibility for digital networks, and yet only a few years after the 1918 influenza pandemic precipitated widespread fever, overheating offered a thermal metaphor to make sense of the spread of broadcast media.[39] As radio became available across the United States, people came into contact not only with domestic radio sets and the voices transmitted through the air but with radio transmitters themselves. Scientists and engineers working with these transmitters began to notice a curious effect: a sensation of heat, "feelings of warmth near radiating tubes."[40] The air around them remained cool, but their bodies feel hot. This was a literal "radio fever."

As radios heated bodies, people began to think of radio as a thermal tool. Radio heat inspired doctors, engineers, and scientists. They wondered how radio's heat rays could be channeled. Instead of a nasty side effect of shortwave transmission, could they be made productive as power, as tools, or as medicine? In Schenectady, New York, at the General Electric Research Laboratory, Willis R. Whitney saw that engineers working with shortwave transmitters, especially in proximity to the shortwave vacuum tube, developed "fevers." Whitney began to experiment with the shortwave tube's heat effects, and by 1928 provided an array of thermal demonstrations to viewers: an apple was placed on a receiving aerial and baked to the core; a sausage, placed in a glass tube, was cooked without fire; electric lamps were lit without wires or sockets. One journalist reported that the vacuum tube produced "a warmth in nearby spectators reminiscent of prohibited stimulants."[41] In the years that followed, the Schenectady laboratory continued to offer demonstrations: spectators witnessed "corn popped by radio," even though the kernels were placed between two glasses of ice water.[42] Radio heat was made public.

Apart from these stunts and exploitations, Whitney saw immense medical possibilities in shortwave heating. In the General Electric Laboratory, Whitney developed the radiotherm, a device that intentionally created an "artificial fever."[43] Others quickly took up the machine for experimentation. Down the road from the General Electric Laboratory, at Albany Medical College, Dr. Helen R. Hosmer used the General Electric transmitter to perform shortwave experiments on animals. Placing tadpoles in liquid between two metal plates connected to the transmitter, Hosmer found that the machine could increase tadpoles' temperature by 3°F in just thirty-one seconds.[44] Another set of scientists attempted to use Whitney's radiotherm to treat polio in monkeys, but after killing several monkeys, they reported inconclusive results. Others directed the fever machines' rays at syphilis-infected rabbits. At the New York State Psychiatric Institute and Hospital, the radiotherm was also used experimentally. In 1930, physicians met in New York to discuss the machine's early results. They agreed that they had to be careful, managing heat in patients not only with the fever machine but also with air temperature, blankets, and body position. If the patient perspired and moisture accumulated on their skin, radio waves could arc to the body through the patient's sweat.

For the physicians and scientists experimenting with radio heat, it offered a modern update to a long history of using heat "as a physical means of alleviating and curing disease."[45] It promised to overcome problems with other heating methods: hot baths and sweatboxing, for example, posed "technical difficulties," and each was seen as working merely on the surface of the skin rather than on the body as a whole.[46] Of much interest was the potential for the radiotherm to replace the use of malaria in producing fevers, which was especially important in the treatment of progressive paralysis caused by syphilis. Physicians imagined that, with a fever machine, they would no longer have to deal with the difficulty of keeping a supply of malarial parasites on hand and alive, nor would they have to maintain the temperature generated by an infection. Heat rays offered a technical solution. Scientists discovered that radio fever could increase the temperature not only of tadpoles, rabbits, and monkeys but also of electrolyte solutions and organ tissues ranging from brains to testicles. If the electrical properties of particular tissues could be assessed, some argued, then this would enable selective heat treatment of one part of the body and not another.[47]

Shortwaves were used both to target pathogens on human bodies and to combat pests. Impressed by the reports on the use of shortwaves for artificial fever treatment, a chief engineer of the Baltimore and Ohio Railroad Company asked if shortwaves could be used to exterminate animal life. Facing massive grain losses because of the grain weevil, J. H. Davis designed an experiment to give the insects a "killing dose" of radio waves by raising the temperature in grain storage beyond their threshold for survival.[48]

At the same time that physicians were experimenting with heating syphilis patients with shortwave radio, surgeons were using a "radio knife" to conduct "virtually bloodless" operations in which they burned, rather than cut, into their patients' bodies.[49] Radio waves allowed them to pierce the skin and to cauterize the wounds as they went, "sealing tissue as with a hot iron."[50] Here, heat rays enabled surgical precision. They also transformed the body into a receiver of thermal communications. One 1931 report stated, "The patient, himself, is the antennae of a miniature broadcasting station when the radio knife is used."[51] As in the case of the artificial fever machine and the air conditioner, the radio knife helped to achieve purification, a body "freed from any possibility of infection."[52]

Even as Whitney's radiotherm and other technologies were developed in the 1920s, high-frequency currents had been used to warm things for years. In the 1890s, Nikola Tesla demonstrated that radio frequency currents could be used to heat human bodies for therapeutic benefits and for surgery, prefiguring the radio fever machines. During this same period, Jacques-Arsène d'Arsonval applied high-frequency currents to the human body and found that they could produce a sensation of warming in the skin. An early version of the radio knife was patented by Lee de Forest, inventor of the Audion vacuum tube, who demonstrated the technique to surgeons in Berlin and Paris in 1908. Electrotherapy—the manipulation of the body through electrical current—had become a new medical field.

In the 1920s, however, it was not just high-frequency currents that were of interest but the radio apparatus and radio transmission as a media form. The spread of broadcast radio catalyzed these experiments with radio heat.

Radio, as a medium, was in turn imagined as a heat transmitter. Some believed that this new media system would create a new form of thermal transmission: broadcast heat. In 1926, the president of the American Society of Heating and Ventilating Engineers, Samuel Edward Dibble, announced, "It is no more improbable to broadcast heat waves than it is to broadcast sound waves."[53] (Notably, just a few years prior, Dibble's university students had formed the bulk of the primary research subjects for the comfort zone chart.) This problem of "sending heat to consumers via the air" was simply a problem of finding appropriate instruments for the control of heat waves.[54] And this was a necessity, Dibble claimed, because fuel stores were gradually being depleted and the future would "see huge centralized heating plants broadcasting heat to homes."[55] These centralized plants would, of course, heat bodies, but they would also free the air of its "impurities."[56] This speculation occurred alongside the thermostatic promises of "leveled heat," the deployment of air-conditioning systems in theaters, the emerging thermocultures of coldsploitation, and the intensification of sweatboxing as a police tactic. It was part of a larger transition to atmospheric thermocultures.

While Dibble's study was never realized as an actual infrastructure, broadcast heat, the mass communication of heat to human bodies via the electromagnetic spectrum and via radio waves, gained traction in the popular imagination. In 1930, one op-ed announced the latest news: "A radio tube may soon replace the furnace, parlor stove and kitchen range for heating purposes. Radio is seen as the heating plant of the future. It will ultimately give you entertainment programmes, political speeches, news events, television pictures, a hot fire and plain home cooking."[57] What was important about "heating by radio" was not simply that it promised, like the thermostat, to reduce all the labor of heating and cooling, but that it would heat the body instead of the room: "Every person becomes an individual range."[58] After overseeing the development of the "fever tube," Willis R. Whitney speculated that radio would also be used to generate body heat in homes, offices, and factories.[59] "Why heat thousands of cubic feet . . . merely to keep a body warm?" he asked. "Maintain sufficient heat in the person, instead of an excess of steam in radiators."[60]

The imagination of sending heat through the air was closely tied to the imagination of broadcasting power. During this period, many claimed that power would finally be sent to distant points through the air.[61] At the 1933 World's Fair in Chicago, Westinghouse mounted an elaborate exhibit featuring the work of radio shortwaves. One review of the exhibit opens, "Sending power through the empty air, long a dream of scientists, was demonstrated recently for the first time."[62] Just as General Electric cooked sausages and remotely powered light bulbs for spectators in its demonstrations, the *Powercasting* exhibit featured a radio antenna that drove an electric motor from thirty feet away. Food was cooked between two electrodes (which remained at room temperature). Spectators wondered at the effects of these heat rays as they stepped into the Powercaster's electric field. This was, one reports, a "radio cocktail" that offered an exhilarating "jag" followed by a "hangover"—a precursor to the Active Denial System, albeit using a different part of the spectrum.[63] Eggs and steak, people imagined, would soon be cooked by radio. As in the other applications of radio heat during the 1920s and 1930s, these demonstrations imagined spectral exchange not only as a means of communication but as a means of heat transmission.

Although heating by radio, whether through a fever tube or home heating system, aligned with the atmospheric transitions of this period, it pioneered a new mode of thermopower, one in which in which temperature was managed spectrally through radiant thermal media. The radical potential of this form of heating existed in the capacity to move thermal transmissions through an environment without heating that environment. The fever machine was not a sweatbox—it was not intended to create a microclimate in which a person could be overheated. It was, like the Active Denial System today, a means of direct targeting and individuation. Even in this early period, scientists were experimenting with directly targeting particular tissues in the body. By the late 1930s, companies such as the Scientific Diathermy Corporation made available portable radiothermy kits for purchase or rental, which offered "soothingly pleasant penetrating radio rays" that brought individual users "almost immediate relief from pain."[64]

At this moment the human body was also conceptualized as a radiator. Studying radiant emissions, scientists found that the human body also emits radiation. One article reported on James D. Hardy's study: "Just as the stars in space radiate away much of their substance in light and heat waves, so the human body dissipates about one-half of its heat and energy daily in waves which can be measured as radio waves can be measured."[65] Hardy estimated that the emissions of a "normal human being" were equivalent to a 40-watt electric bulb—and fell within the infrared spectrum, "between the extremely long radio waves and the unusually short rays of visible light."[66] Here, too, heat was located between radio and light, and bodies were seen not simply as receivers of radiant waves but as heat emitters. Hardy believed that the feelings of heat and cold could therefore be reduced to improperly balanced emissions. In the summer, because there is excessive heat, humans don't radiate, Hardy hypothesized. And in the winter, because of the cold, there is too much radiation emitted. The correct amount of radiation, the study claimed, would produce comfort.

To return to the question of radiance: heat had always been spectral. It had always been radiant. Even before Herschel's discovery of infrared, people developed thermal media, such as clothing, to manipulate radiation—the radiation of their own bodies, of fires, and of the sun. Even without a technical understanding of radiation, enslavers were able to manipulate solar radiation to enact thermal violence. Thermal media doesn't depend on technical or scientific knowledge. But in the early twentieth century, the focus on spectral forms such as projected light and radio waves prompted the development of new media technologies that were designed explicitly to channel thermal transmissions. People tinkered technically and scientifically to generate thermal effects. Heat was approached as a spectral phenomenon and directed as a particular kind of communication: the heat ray.

The imagination and practice of thermal manipulation—whether in the process of medical operations, cooking, or home heating—was entangled with the materiality and sensibility of media. Heat rays emerged out of experiments with communications systems and took shape in relation to media systems. As a result, radiant heat, infrared signals, and fever tubes were all inflected by the ideals, operations, and forms of wireless media systems and communications devices. Media cultures, and especially the cultures of telegraphy and radio transmission, were enfolded into thermocultural practices that situated bodies and food as the direct receivers of heat transmissions. Radio works, critically, as a medium for communicating at a distance. Radiant heating is not a convective process. It is not, like the thermocultural atmospherics of coldsploitation, oriented toward the creation of an environment. It is not designed to transmit heat into air. In these early twentieth-century experiments with infrared and shortwaves, the thermal environment was one of neither immersion nor direct contact; it was heating from a distance without appearing to affect the environment in between the sender and the receiver. Moreover, these experiments reveal that in the early history of our communications systems, it was neither clear nor necessary that the substance of mass media would be light or sound. Radio was not simply a medium of sound; it was conceptualized as—and became, in medicine, for example—a medium of heating. Telephony was a medium not simply of electricity but of infrared radiation. And even as the spectral experiments during this period formed the basis for broadcast systems, they also established the material foundation for many personalized thermal systems—technologies that could target individual bodies via radiant thermal media.

THE INFRARED INTERNET

The early experiments with optical communications and infrared light encountered a number of challenges that limited their capacity for media circulation. They faced the problem of interference: atmospheric effects could easily disrupt their transmissions. A cloudy day could inhibit the use of sunlight as a medium. Even Alexander Graham Bell's photophone experiments were troubled by the weather, and artificial light was brought in as a substitute.[67] Infrared ends up being limited in its capacity to transmit through fog. Frustrated by the limits of infrared in the Noctovisor, John Logie Baird ultimately substituted in a radio wave transmitter. At the same time, inventors faced the problem of the availability of a viable heat transmitter. Depending on the sun meant depending on the weather. In experiments during World War I, a carbon arc lamp was the most powerful medium available, but it placed severe limits on both the quality and the distance of the transmission. And a problem existed in the effects of both light and heat: Visible light produced blinding effects. Infrared waves overheated their subjects. For all these reasons, infrared radiation was not taken up as the substrate for mass communications. It was eclipsed by wired technologies and wireless transmissions using other parts of the spectrum. Radio waves became a national medium. Radar was used to target enemies. Television signals crisscrossed the globe.

Heat rays, like the coldsploitation media of this same period, still formed a foundation for a range of twentieth-century thermal media. The shortwaves used to cook sandwiches were domesticated in home microwaves. The targeted beams of the heat ray gun were realized in the Active Denial System. The investigation of the spectral properties of matter expanded far beyond the capacities of the spectrophone or the spectroscope, and spectral analysis became a foundational part of astronomic research, among many other scientific fields. As I explain in chapter 5, infrared photography became fully realized as a sensing technology. And eventually, remote controls were developed to communicate with television sets using waves in the near-infrared band, typically around 940–980 nm. Despite the fact that the heat ray experiments inspired so many developments, the 1881 *Scientific American* assessment—"Whatever be the future before the photophone, it assuredly deserves to rank in estimation beside now familiar names of the telephone and the phonograph"—did not come to pass.[68] The photophone is largely considered a failure by media historians, a "clumsy looking machine that did not work."[69]

Yet many people in the field of fiber optics today—engineers developing the technological infrastructure for digital communications—hail the photophone as "the most significant optical communication system development."[70] Its invention is marked as a significant "historic event" in the first pages of optical communications textbooks.[71] As optical engineers recount their history, the photophone first demonstrated the core principle of communication via light that would later pervade all optical networks. It failed, they argue, because of the lack of a dependable medium for transmission and the lack of a reliable light source. They point out that it was not until the 1960s, with the advent of a better medium for optical communications, that the historical progress toward the creation of fiber optics could resume.

In this history—foundational to fiber optics but less often described in the pages of media history books—the development of the laser, an intense beam of a single wavelength, solved the problem of needing a reliable light source. Given that light waves have a higher frequency than radio waves, they can carry ten thousand times more information than even the highest radio wavelengths.[72] Bell Labs' communications engineers began to work on a series of theoretical calculations for laser communication in the late 1950s, but it was not until 1960 that the first operational laser was developed at Hughes Research Laboratories in California. Unlike radio waves, lasers were "unsuited for open-air transmission" precisely because they were "adversely affected by environmental conditions," including such conditions as rain, snow, and smog.[73] The problem still existed: radiant light, visible and infrared, still needed a reliable transmissions medium.

In the history of optical communications, the answer to this problem was glass. In 1966, two researchers at Standard Telecommunications Laboratories in England proposed that glass—optical fiber—could be a suitable transmissions medium. The outstanding problem, they pointed out, remained one of attenuation. They proposed that the signal needed to attenuate at less than 20 dB/km. If that threshold could be met, lasers could be sent down optical fibers. This catalyzed attention to purifying glass: the more homogeneous the glass, the less attenuation would occur. Much of this experimentation happened at Corning, located in upstate New York. In the 1970s, researchers there developed a glass fiber, "the purest glass ever made," that met the proposed threshold.[74]

As was true in many prior optical experiments, there was a tacking back and forth between visible and infrared wavelengths. At Bell Labs, engineers experimented with infrared wavelengths before moving to visible light. When Ted Maiman generated the first laser in 1960, it consisted of visible light, a deep ruby—at 694.3 nm. The same year, a group of three other physicists created a laser in the infrared band. And two years later, another engineer developed a laser diode that emitted at 850 nm. It was in this wavelength, in the near-infrared band, that the earliest fiber-optic systems were produced. This wavelength was selected in part because technologies for light emission and detection (including less-expensive silicon detectors), produced for visible light LEDs, could be used. It was also made possible by the sensitivity of a silicon cell (rather than a selenium cell), which extended further into the infrared spectrum (1170 nm) with a peak between 800 nm and 900 nm (also in the infrared range). As was true for so many previous infrared technologies, infrared was adopted for early lasers in part because of its adjacency to the visible part of the spectrum.

However, the selection of the infrared band of 850 nm—and later 1310 nm and 1550 nm—also reflects the material constraints and potential elemental interference within silica-based fiber. On the one hand, fibers are not entirely homogenous. They contain minute fluctuations in glass composition and in density. As radiation travels down the fiber, there are "elastic collisions" between the waves and the silica molecules that result in a process called Rayleigh scattering.[75] If the scattered light "maintains an angle that supports forward travel within the core," then the signal doesn't attenuate.[76] Otherwise, the light is diverted out of the fiber core, and the strength of the signal drops. In short, this means that visible light, which has shorter wavelengths than infrared, attenuates more quickly as it is funneled down a fiber.[77] On the other hand, when longer wavelengths are used, especially beyond 1550 nm, "ambient temperature becomes background noise, disturbing signals."[78] In other words, the longer the wavelength in the infrared range, the more heat can interfere. Moreover, within this zone, between the visible and the long infrared, there are "absorption peaks," bands caused by water vapor collection in the glass, which also cause attenuation at specific wavelengths (figures 4.3 and 4.4).[79] No matter how pure these fibers are made, they can never be entirely homogenous or insulated from the outside world. Thermal entanglement continues to inflect communications technologies.

Early experiments with infrared fiber transmissions thus began with the 850 nm wavelength, which corresponds to a window in the silica-based

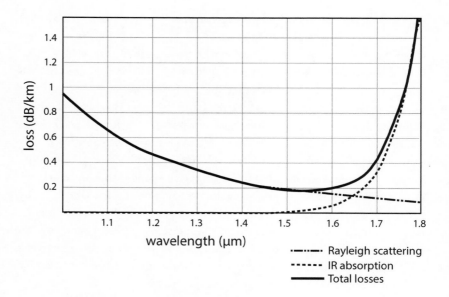

FIGURE 4.3. This diagram reveals how Rayleigh scattering (the conflicts caused within a fiber) increases at shorter wavelengths while infrared absorption increases at longer wavelengths. From "Optical Fiber Loss and Attenuation," FOSCO: Fiber Optics for Sale Co., January 27, 2020, https://www.fiberoptics4sale.com/blogs/archive-posts/95048006-optical-fiber-loss-and-attenuation.

fiber where there is less loss of signal—and, as a result, less attenuation.[80] But this still has a relatively high loss limit of 3 dB/km. Engineers then jumped to a second infrared window at 1310 nm (after a competition with 1060 nm), which dropped the attenuation to around 0.5 dB/km. And by 1977, the Japanese telecommunications company Nippon Telephone and Telegraph developed a third infrared window at 1550 nm, bringing attenuation to about 0.2 dB/km. As one optical communications textbook synopsizes, "Each wavelength has its advantage. Longer wavelengths offer higher performance, but always come with higher cost."[81] For shorter transmissions lengths, more loss is acceptable, and thus the wavelengths of 660 nm and 850 nm are used, but the "longest link lengths require 1550 nm wavelength systems."[82]

Today, these three wavelengths of infrared radiation—850 nm, 1310 nm, and 1550 nm—have been adopted across digital communications systems. The National Institute of Standards and Technology "provides power meter calibration at these three wavelengths for fiber optics."[83] More

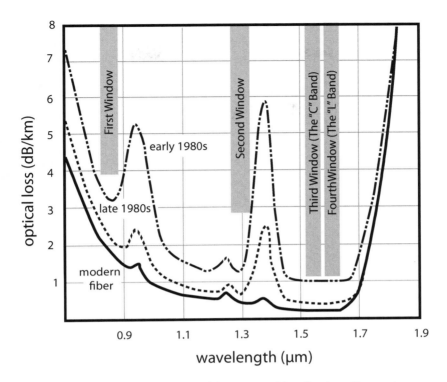

FIGURE 4.4. In this diagram, produced for a practical handbook on fiber optics, the authors show correlations between optical loss and wavelength. From David R. Goff and Kimberly S. Hansen, *Fiber Optic Reference Guide: A Practical Guide to Communications Technology* (Burlington, MA: Focal Press 2013), 4.

recent updates in fiber-optic transmission, such as wavelength-division multiplexing, have expanded the range of wavelengths that can be used—typically in bands that range from 1260 nm to 1675 nm—but all these remain in the infrared spectrum. And beyond the fiber-optic network, infrared lasers, like the heat rays before them, are deployed in scientific, industrial, and institutional contexts. Lasers operating at 1064 nm and 10,600 nm, both in the infrared region, are used in surgery and industrial laser cutters. Lidar systems, which pulse laser beams to measure the distance to an object and to image the world, often use 905 nm and 1550 nm wavelengths. These technologies, all of which entail creating directional rays using the infrared spectrum, have become essential not only to the operations of digital media but to the construction of autonomous vehicles and contemporary sensing technologies. These, too, face the problem of atmospheric conditions, including fog, clouds, rain, and snow, which make it difficult to direct and receive infrared rays.

In digital media theory and in popular representations of the internet's infrastructure, its substance is often a blinding visual light. In digital networks, Wendy Chun writes, "rays of light serve as communicative fibers between all human beings."[84] Jeff Hecht's history of fiber optics, *City of Light*, begins with early experiments to guide visible light along jets of water.[85] The canonical image of a fiber-optic cable depicts illuminated glass strands. In artworks such as *Filament Mind*, fiber-optic cables are illuminated by pulsing, multicolored lights (figure 4.5).[86] Those who study and write about fiber-optic infrastructure as a medium of light are not technically incorrect. In the technical descriptions of the spectrum, infrared transmissions (along with ultraviolet transmissions) are described as light—though it is important to note that they are a form of *invisible* light. Counter to the dictionary definition of *light*, which says it is "something that makes vision possible," the optical communications field defines *light* neither in relation to sight nor human vision.[87]

This slippage—in which fiber optics transport invisible light but are described as transporting light, which is normatively equated to vision—

FIGURE 4.5. Brian Brush and Yong Ju Lee's *Filament Mind* (2013), installed in the Teton County Library in Jackson, Wyoming, depicts visible light transmitted through fiber-optic cables. The cable artwork illuminates terms Wyoming library patrons have searched for. From yongjulee, "Filament Mind," February 13, 2013, accessed April 5, 2021, https://vimeo.com/59626512.

obscures the critical importance of the infrared spectrum to digital communications today. The substance and medium of digital communications are not visible, nor are they part of the electromagnetic spectrum to which human eyes are acclimated. Rather, the future of transmissions development lies in moving farther into the infrared—a fourth window at 1625 nm is being developed, although temperature generates more difficulties in this band. While the development of early visible light technologies (such as the LED) shaped the move into the infrared (at 850 nm), today technical development is focused on the invisible bands of the near-infrared, such that fiber-optic development supports growth in the areas of autonomous driving and infrared-based lidar sensing. Our information is literally transported around the world not via beams of visual light but by the heat rays anticipated by nineteenth- and early twentieth-century inventors. The internet is comprised of radiant media and thermal circuits.

THE CABLE THERMOMETER

November 2011, Antarctica. Just a month before an all-time high temperature was recorded at the South Pole, researchers on the McMurdo Ice Shelf drilled two boreholes. They established two long, thin openings linking the ice surface to the water beneath. Into their ice conduit they lowered two fiber-optic cables. These quickly froze in place. This is a communication system—not for sending messages to the ocean but for retrieving temperature readings. The fiber-optic technology, called distributed temperature sensing, offered something that glaciologists previously lacked. Before, their instruments needed to be accessed, powered, sent, and retrieved. And even with the array of remote sensing devices available—satellite, airborne laser scanning, and infrared imaging—the zone beneath the shelf evaded continuous capture. With the direct cables, they now had the ability to consistently capture the temperature of water beneath the shelf. They could generate high-resolution data about the waters that warmed the ice from below. This kind of data capture, they suggested, might offer a more direct measurement of the changing ice sheet.[88] These systems could capture areas that remained inaccessible to aircraft or satellite. They could finally retrieve data from a single site over time.

The fiber-optic cable's thermal sensitivity and its capacity to act as a thermometer are spectral phenomena. As infrared waves travel down a glass cable, they scatter. Redirected at slightly different wavelengths, some of these waves have "a unique temperature dependence."[89] By reading the backscatter of these waves, engineers are able to read thermal changes along the cable route. When they calculate these changes, the fiber-optic cable becomes "a very long continuous thermometer."[90] The cable and its infrared transmissions can be used to register an array of activities in the environment around it. Since all activities have a "thermal signature," this turns the cable into not only a long and continuous thermometer but also a sensing and eventually a surveillance mechanism for the world around it.

The use of cables as long and continuous thermometers was pioneered not by environmentalists invested in documenting climate change but by the extractive industries. When fiber-optic systems and distributed sensing were developed in the 1980s, engineers imagined that they held immense potential for industrial purposes.[91] In the power industry, they could be placed near high-powered conductors without disruption (unlike copper cables). In power plants, they could help optimize electrical flow and locate hot spots.[92] By providing continuous monitoring capabilities in harsh natural and industrial landscapes, these engineers imagined that fiber-optic sensing might make infrastructure systems—power supply, water distribution, and manufacturing, among others—more secure. Manufactured out of the same material as internet cables, these fiber thermometers enable the remote surveillance of the extraction of fuel from the environment. In "smart fields," they relay temperature changes inside oil wells and can be used to assess the effectiveness of valves. Slight shifts in temperature are used to determine the flow of different fluids (each of which has its own thermal behavior). In 2011—just as the McMurdo cables were being installed—Roctest, a company specializing in geotechnical sensors, extended a line of fiber-optic cable at the world's largest copper mine, in Chile's Atacama Desert. The cable, several kilometers long, became part of a monitoring system that registers "abnormal infiltrations" of liquid acid through a dam.[93] Because temperature-sensing cables can detect leaks, they are at times laid next to oil and gas pipelines. In mines, they are used to monitor for ventilation problems and potential explosions, and in underground natural-gas storage facilities, they register available capacity. Thermal variations are harnessed to make the extractive industries more productive. Temperature is the language, remediated through optical cables, through which security can be achieved.

The distributed temperature sensing system on McMurdo Ice Shelf, however, was part of a more recent attempt, especially by glaciologists and hydrologists, to use fiber-optic cables to record the impacts of climate change. The adoption of such cables, scientists claim, "opens a completely new window on river dynamics."[94] Distributed temperature sensing not only uses the cable to index temperature; it also uses temperature to document an array of environmental changes, such as glacial melt. Scientists then began to wonder if the vast amount of cable stretching under the oceans, the subsea cables that carry 99 percent of all transoceanic internet traffic, might also be transformed into sensors. Marine scientists in particular, who needed to collect data from and about the deep ocean, saw a huge opportunity to better understand the ocean itself, seismic events, and climate change.

Scientists had long used cables to link to stand-alone undersea sensors. In the 1990s, sensors were attached to the end of retired undersea cables for earthquake monitoring off the coast of Japan. In 2006, fiber-optic systems were used for the first dedicated "cable observatory" off the coast of British Columbia. In a cable observatory, custom-laid fiber-optic cables are intended not to transport telecommunications traffic but to bring back information about temperature, pressure, currents, and seafloor activity from sensors on the seafloor. And in 2011, following the Tohoku earthquake, Japan deployed a large array of pressure and temperature sensors as a means of environmental monitoring.

By 2012, a joint task force had been developed by three UN agencies to examine the possibilities for developing SMART (science monitoring and reliable telecommunications) cables. In the early proposals, the task force advocated for technologies that could be incorporated into telecommunications cables' repeaters to measure temperature, pressure, and seismic acceleration. As the group argued, "The market-driven investment in information infrastructure can be harnessed to achieve tangible, social benefits in climate and ocean science."[95] These hybrid science-telecom cables face challenges, however. As they transform "neutral" communications circuits into environment-sensing devices, they also invoke the possibility that these systems could be used for nonscientific purposes, such as military monitoring, which might infringe on territorial sovereignty in the coastal oceans.

At the same time, the scientific community began to wonder if, instead of placing temperature sensors in the cables' repeaters, they could use distributed temperature-sensing equipment on the internet as a whole. In 2018 and 2019, a number of studies were published that demonstrated the feasibility of using dark fiber—the unlit, open capacity on existing telecommunications cables for internet traffic—for geological sensing. These studies used techniques from previous distributed temperature sensing systems, especially changes in Rayleigh scattering, to sense the world around these cable systems (a process called "optical time domain reflectometry").

Scientists collected seismic data on existing cables between West Sacramento and Woodland, California.[96] Others gathered data from Iceland's Reykjanes Peninsula that enabled them to view volcanic dikes in unprecedented resolution.[97] And using undersea cables off the coast of Monterey, California, scientists demonstrated that the same principle worked underwater: the cable could simultaneously transmit data between end users and information about subsea movements.[98]

These studies suggest that the vast networks of telecommunications cables that materially support the internet can be transformed into a massive earth sensor. Temperature could be read from more than 1.1 million kilometers of internet transmission lines running along the seafloor, which encircle the globe and stretch to the deepest parts of the ocean. This is all possible because fiber-optic cables are radiant media that operate by channeling infrared radiation down the substrate of glass. While radiant transmissions' propensity for interference, disruption, and scattering is one of the key factors that kept the photophone, heat ray telegraph, and other infrared transmissions from being adopted in the early twentieth century, this tendency to intra-act with the "outside" world, this ongoing thermal entanglement, is what could enable the backbone of the internet to become the world's longest thermometer.

SPECTRAL THERMOPOWER

The heat ray, as I describe it in these pages, is a particular kind of radiant thermal technology. It is not, like some architecture, a means of manipulating materials to concentrate or deflect the sun's thermal radiation. It is not, like clothing, a means of trapping thermal radiation from the body to create an inhabitable microclimate. While these are instances of radiant mediation, they largely operate to modulate atmospheres. As a result, their management of temperature works through the production of zones, environments, and climates. People can be controlled by being placed in iceboxes and sweatboxes, by stimulating attractions to particular climatic forms, and by creating interfaces that separate bodies from environments. Experiments with the heat ray have operated somewhat differently. Heat ray technologies have approached infrared radiation and thermal sensation as spectral forms that can be channeled through thermal circuits. It was during the late nineteenth century and early twentieth century that infrared was envisioned as a communications medium for information flows and radio became a heat-delivery mechanism.

But these early histories also reveal the difficulty of working in this part of the spectrum. These waves scatter when they come into contact with clouds, rain, and other environmental phenomena. Just as Baird's floodlight television blinded its subjects, so did its hot plates overheat them. Because radio transmissions, and their incredibly long waves, interact less with human bodies and the environment, they are more viable forms of communication. And even then, shortwave radio's unexpected production of heat effects generated new imaginations for how human bodies could be managed as a spectral form. Discussions about spectrum scarcity and regulation, Helga Tawil-Souri shows, often visualize it as a contested territory, and the zone of radiant thermal media—as are others—highly populated and perpetually occupied.[99] This materialized most recently in conflicts between 5G networks that proposed to send cell signals using the 23.8-gigahertz frequency and weather forecasting systems that use this same frequency to detect the radiant emissions of water vapor.[100] Spectral media are disrupted by weather and in turn disrupt climate sensing, and yet their scattering—which occurs even in the core of the fiber-optic systems that transport the vast majority of internet traffic—also makes possible thermal sensing along the internet's base layer. As fiber-optic cables, and all their unlit capacity, are envisioned as a potential sensor and a thermometer, they also promise, like the cables in both science and in the extractive industries, a new form of monitoring and surveillance.

This chapter reveals how thermopower operates not only through the calibration of subjects, the management of atmospheres, and the direct manipulation of bodies as media. It also exists in the capacity to harness and channel infrared transmissions via radiant thermal media and to produce heat effects through spectral modulation. The increasing investment in this part of the spectrum and on the production of thermal effects through the spectrum is generating more capacities to manipulate heat in and through spectral processes. What radiant thermal media offer is also of critical importance in the current moment and in the context of climate change: the transmission of heat, and power, at a distance, and the capacity to isolate and discreetly target individuals for thermal transmissions.

5

INFRARED CAMERA
THE THERMAL VISION OF HEAT IMAGES

Although heat rays are often described as invisible, people are able to see temperature in its many radiant, convective, and conductive forms. One can witness heat effects even without thermometers and weather channels. Fevers produce sweats. Chills produce goosebumps. Leaves change color and fall. Ice crystallizes and water evaporates. Condensation on a window signifies a thermal difference between inside and outside. Towering cumulus clouds on the horizon foreshadow a thermal shift. Animals gather in the shade. Movement indicates that beings are hotter or colder than their surroundings, losing heat or gaining heat, or undergoing phase transitions. Thermal entanglement, the inevitable intra-actions between matter and its thermal contexts, leaves a multitude of visible traces.

The term *thermal vision* describes the ways of seeing these thermal exchanges. People learn the visual grammar of thermocultures: While cooking, they look for boiling water (a sign of crossing a thermal threshold). While managing livestock, they look for lethargy and shivering in the animals (signs of thermal stress). When fighting fires, they look for puffing smoke (a sign of backdraft). Sometimes these signs are concretized in the visual codes of a thermometer or a heat map. Even in the absence of actual thermal manipulation, however, visual media shape the meanings of temperature.[1] Cinematic images of a bright sun and heat haze tell spectators that it is a warm day. Characters shiver and press up against one another, making sure that their heat does not get lost but rather is transmitted to one another. Thermal vision is synesthetic, relayed through what Laura Marks calls "haptic images"—visual perception extends the sense of touch.[2]

Thermal vision is widely accepted as having embodied temperature effects. In the 1920s, directors, exhibitors, and audiences of coldsploitation described images of snow as conveying actual cool sensations. Half a century later, while making *Do the Right Thing* (1989), Spike Lee reflected on the cinematographer's approach: "He's already thinking about how to visualize the heat. He wants to see people in the theaters sweating as they watch the film. We should have closeups of people's faces, I mean extreme closeups, with beads of perspiration dropping off. Every character must comment on the heat. Those outside should look up at the sky, the sun."[3] The synesthetic effects of thermal vision are harnessed not only for media attractions but also broadly across architecture and design. Color, especially red and blue, has been a long-standing way to both represent temperature and produce thermal effects. Temperature is linked to hue even in ancient color theory. In the twentieth century, psychologists began to test the "hue-heat hypothesis," positing that the color of an object or environment affects the human perception of temperature. Researchers placed test subjects in rooms illuminated with colored lights, gave subjects goggles with color filters, and altered the color of virtual objects in head-mounted displays.[4] Observing their subjects' perceptions of temperature, researchers found that people believe that they are warmer in red rooms, that they feel that drinks are colder in blue cups, and that color functions as a widespread ambient thermoceptive cue.[5] Thermal vision, whether communicated through screens or buildings, is an integral part of how people conceptualize and feel temperature—it is part of their thermoceptive apparatus.

While directors and designers may intend to transmit a sense of temperature, and these images may reflect broader thermal ideologies, thermal vision is a form of perception rooted in embodied experience. Thermal vision elicits experiences of previous thermal exposures and activates thermoceptive regimes that saturate a viewer's body. The effects of thermal images are refracted through bodily history and composition. For a woman in the thermostat wars, the temperature setpoint might be viscerally linked to a sense of control over her body. That same number, for a man who turns down the thermostat during fuel shortages in the United States, might evoke a masculine patriotism. Moreover, the work of thermal representations extends far beyond the screen or the interface. Images of heat and cold, like all thermal descriptions, naturalize thermocultural assumptions and contribute to the long-term calibration of thermal subjects. These images are vectors of thermopower—they synesthetically cement felt connections between red and hot, isolation and cold, masculinity and sweat. While the materiality of temperature shapes imaging and sense-making, none of these links is either neutral or natural.

This chapter traces how a single thermal medium, the infrared camera, has generated a new form of thermal vision, one that diverges from the visual indicators of everyday thermocultures. While many ways of seeing temperature situate it as an invisible force, an environmental property, or a set of conditions to be inhabited, the infrared image casts the world as a landscape of infrared reflectors and infrared emitters—as a field of thermal communication. The infrared image is accompanied by a new thermoceptive regime that indexes bodies, objects, and architectures through these heat emissions. Its deployment generates new "sensing practices" that, as Jennifer Gabrys explains, are modes of sensation, not molded by a universal human reference point but emerging in relation to technologies and environments.[6] Put simply, the infrared camera is not just another thermal medium alongside thermostats, sweatboxes, and heat ray guns: it is a technology whose sensing capacities work to transform all matter, whether bodies or buildings, into thermal media itself.

The first part of this chapter documents the emergence of the infrared camera, a radiant thermal medium, in the early twentieth century. At the same time as scientists were consolidating infrared rays into directed beams and conceptualizing heat as a transmissions medium, the infrared camera generated a new vision of the world. It operated much like a visible-light camera, but it captured spectral reflections in the near-infrared range, just beyond the edge of visible light. In part due to the development of commercial infrared plates in the 1930s, infrared photography was then diffused throughout social and scientific analysis, in areas from medical imaging to audience reception studies. Although, like the heat ray, the infrared camera did not become a mass media form, its social and scientific applications leveraged it to manipulate bodies, objects, and environments. Crucially, the infrared image revealed new ways that bodies, objects, and environments might themselves become spectral media, especially thermal media.

While early infrared cameras remained important in medical, environmental, and military contexts through the mid-twentieth century, a critical origin point for contemporary infrared imaging occurred in the 1970s when the camera underwent a key technical and social transition. In this moment, the infrared image extended beyond portraying a landscape of infrared reflectors and transformed the world into a population of heat emitters. In part driven by the energy crisis of this decade, these new thermographic images were deployed widely as a means of assessing building insulation and as a mechanism of thermal containment. This environmental imperative catalyzed not only developments in the field of thermal imaging but the formation of a company, FLIR Systems, which remains one of the most significant players in the infrared camera market today. The thermography of the 1970s solidified a thermoceptive regime in which the landscape itself came to be seen as a potential thermal medium, open for management and for manipulation. The infrared camera functions as meta-thermal medium.

In the twenty-first century and especially since the beginning of the COVID-19 pandemic, infrared imaging has become integral to an array of social forms. It is woven throughout military and security systems. Its high contrast enables the detection of edges and outlines for artificial intelligence. And its capacity to determine body temperature makes it indispensable to pandemic management. The second section of this chapter dives into two infrared-imaging practices to show how heat images, rendered through digital processing, are used to intensify the management of the environment and subjects as thermal media. In the first of these, I describe how thermal imaging is helping to transition agriculture from broadcast methods to a narrowcast form of precision management, where particular parts of the field are targeted for resource distribution and thermal modification. I then turn to the ways that infrared has gained traction as an animal management tool in wildlife conservation, particularly in Kenya and South Africa, in part due to funding by global environmental organizations. Here, infrared cameras are, like building thermography and precision agriculture, described as an eco-economic tool, but in practice they are used to regulate and reify existing apartheid infrastructures. In each of these examples, the infrared image extends the reach of thermal mediation: into the field, into the bodies of humans and nonhumans, and into social formations. It is a key tool of what Lisa Parks calls the "radiographic episteme," the process of making radiation into "data that can be made productive within an information economy" and a key medium of contemporary thermopower.[7]

Despite the fact that infrared imaging has been used to target, surveil, and manage bodies and populations, it is often seen as a neutral representation: by registering heat rather than light, proponents of infrared imaging claim, it offers a form of vision indifferent to bodily difference and to race, gender, and ethnicity. This chapter's final section turns to the artwork of Richard Mosse, who uses a military-grade thermal weapons system to capture images of refugees. These heat images are a prime example of infrared's claim to neutrality—Mosse claims that they offer a color-blind thermal vision, a portrayal of what Giorgio Agamben calls "bare life."[8] Analyzing these images, I show that although the infrared image promises a more "objective" form of vision, it naturalizes a sense of certain bodies as mere heat and amplifies racialized and sexualized forms of power. Just as metaphors of temperature invoke thermal objectivity to mask racial biases, so, too, does the thermoceptive regime relayed by infrared images.

The story of infrared imaging is often narrated as a history of militarized *Predator* vision, drawing from the 1987 film *Predator*'s imagination of an extraterrestrial creature that hunts humans using its infrared sight. In *Predator* vision, a human body becomes a "spectral suspect," a bright spot cast against a dark background, targeted as a function of machine vision.[9] Infrared imaging has been and remains integral to military vision and practice, a genealogy outside the scope of this chapter but well documented in works such as Paul Virilio's *War and Cinema*. In many places, the thermal camera operates as an extension of the gun and thermal signatures are a means of marking and managing populations.

This chapter's genealogy of infrared cameras reveals a parallel history that runs alongside and informs infrared targeting and surveillance—one in which the infrared camera is intimately connected to modes of environmental and bodily monitoring and management, technologically, industrially, and epistemologically. These images scaffold a thermoceptive regime that purports to convey a new form of objectivity while layering thermal difference into a framework of power, recoding it into "normal" and "abnormal" temperatures. For the viewers of infrared images, heat becomes a form of communication—delivered from the environment, through the camera, to the viewer. The world, in these images, is supposed to be felt not as an external force but a thermal data set with a normative state. Critically, the thermal vision of the infrared camera relays an assumption that bodily and environmental temperature can and should be visually controlled. Capturing an entity in infrared is a means of subjecting it to thermopower.

INFRARED: FROM REFLECTIVE TO EMISSIVE IMAGES

Alongside experiments with spectral transmission, the late nineteenth and early twentieth centuries generated technologies for sensing thermal radiation. Samuel Pierpont Langley, better known for aviation and astronomic experimentation, invented a simple instrument, a "measurer of radiations" that registered infrared energy.[10] The bolometer was refined, and by Langley's 1901 demonstration, it was capable of detecting a cow from a quarter mile away. By measuring temperature at a distance, the bolometer enabled the remote sensing of what had largely been determined from close up: heat, especially heat emitted from living beings. Of course, beings had always been radiant. This radiance had always been sensed. What the bolometer did was extend thermoception from relays between body, skin, air, and other physical transmission media into a standardized technical apparatus.

Like the bolometer, the infrared camera could also detect radiation from a distance. While infrared images were developed in the nineteenth century, it was not until the early twentieth century that they began to be widely circulated. Capturing radiant reflections in the near-infrared region, these early infrared cameras did not register thermal emissions themselves— they did not offer a direct index of temperature, as the bolometer did. Rather, they were able to "see" because infrared radiation from the sun or another source was either reflected or absorbed by a set of objects in view. Addressing the Royal Photographic Society in 1910, Robert Williams Wood asked his audience to consider "how the world would appear to us if our eyes were sensitive to some region of the spectrum other than the one for which they have become adapted."[11] In his lecture, Wood presented a series of infrared photographs, including images of trees, whose leaves are as bright "as if colored by freshly fallen snow" and sheets of white paper that appear dark, even set against a blue sky in full sunlight (figure 5.1).[12] These images opened the infrared spectrum and its "invisible rays" to visible perception and, as Wood argues, reveal "the fact that we do not see things as they are."[13] These images offer a vision not of body heat but of the world's capacity to transmit—reflect, deflect, or absorb—infrared rays.

The release of infrared-sensitive plates by Eastman Kodak in the early 1930s made infrared photography much more accessible and catalyzed its expansion as a medium. In September 1932, the seventy-seventh annual exhibition of the Royal Photographic Society in London offered views of the invisible: Infrared photographs penetrated haze. They captured hot irons, even in complete darkness. And they revealed passages blocked out by the

FIGURE 5.1. "A Landscape by Infra-Red Light," by Robert Williams Wood (1910). From Wood, "Photography by Invisible Rays."

censor of the Spanish Inquisition in rare and old books.[14] In the typical fashion of new media, the infrared craze promised to transform all cultural forms. In a short primer on infrared photography released in 1933, S. O. Rawling recounts the "spectacular manner" in which infrared photographs had emerged before the public in the previous three years. "The impression is now prevalent," he writes, "that a fundamentally new discovery in photographic science has recently been made."[15] But this was not so, as he reminds his readers of the early twentieth-century experiments, such as Wood's photography. Alongside a wealth of reflections in the popular magazines and newspapers of the time, Rawling saw infrared imaging as the next frontier of photography—a new form of thermal attraction. In 1932, *Popular Mechanics* reports that thanks to the Eastman plates, "infra-red photography is now possible to the amateur."[16] Describing the new plates in *Camera Craft*, one author jokes, "You can take a picture of black cats in dark cellars if the cats are there providing the felines are Hot Stuff."[17]

Although infrared was not taken up widely by amateurs, the release of these plates and the commercialization of infrared technology dispersed infrared photography throughout numerous scientific, technical, and industrial practices. Prior to the release of the plates, infrared imagery had already extended into a number of these contexts, from astronomy to medicine. Sometimes called "thermograms," infrared images were often used as a means of detecting heat and measuring a subject through its radiant reflections. The technical developments of the 1930s broadly expanded the social and scientific potential of thermal vision. Astronomers used the plates to detect nebulae and clusters. Scientists used infrared plates to detect plant diseases (figure 5.2).[18] Anthropologists looked at infrared photographs of Black, Japanese, and white men and tried to parse what these

FIGURE 5.2. An infrared image of a potato leaf reveals lesions. From R. N. Salaman and C. O'Connor, "A New Potato Epidemic in Great Britain," *Nature* 134 (December 1934): 932.

images revealed about race.[19] And infrared photography spread across the medical professions to be used in the detection of circulatory problems and the diagnosis of tumors.[20] In the media industries, infrared photography was used to gauge audience reactions, dovetailing with cinema's industrial expansion.[21]

Many of these imaging practices, which permeated medical thermography in particular, remapped bodies and objects as a set of gradients. They produced a series of proto–heat maps that, even if they did not directly indicate temperature, cast the landscape into healthy and diseased areas, with correct and incorrect temperatures, almost always with "incorrect" infrared absorption and reflection as an indication of a problem. For example, in early infrared images of patients with heart disease, veins form a dark spiderweb, a set of contoured indications of a hidden circulatory system, visible only in its distention (figure 5.3). Images of the nonhuman world

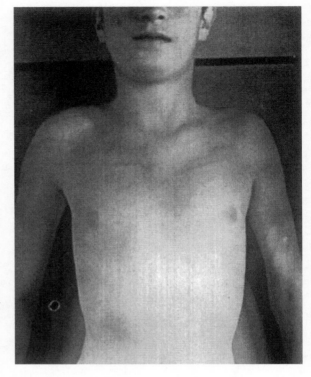

FIGURE 5.3. An infrared image reveals the enlarged veins just beneath a boy's skin. From B. S. Epstein, "Infrared Photographic Demonstration of the Superficial Venous Pattern in Congenital Heart Disease with Cyanosis," *American Heart Journal* 18, no. 3 (1939): 285.

also adopt this orientation. In the image of a potato leaf (figure 5.2), the leaf is similarly separated into light and dark zones, with the dark regions indicating lesions of potato blight.

These images permeated numerous medical research fields in the subsequent decades, and as Lisa Cartwright, José van Dijck, and others trace, they supported not only scientific diagnostic practices but a disciplinary and managerial gaze that governed the body.[22] What became visible, whether blight or circulatory system, was visible because of thermal difference, and in the process it was coded as abnormal. Infrared imaging not only revealed the invisible but asserted and made visible a set of norms and deviations, drawn into relief in light and dark, and implied a set of corrective procedures. The subjects of these images became new subjects of medical and ecological management: they were not only indexed through their thermal reflections but cast as sites for thermal manipulation—from therapeutic regimes using heat and cold to temperature management in the growing field.

It is not surprising that infrared photography was seized as a means of imaging and imagining the world in the 1930s. During this period, the invisible rays of radio were being carried across the electromagnetic spectrum, into people's living rooms, social spheres, and imaginations, and infrared more broadly had become a site of intense experimentation. Infrared photography provided a view of these invisible emissions—but critically, it offered a view of bodies' and objects' capacity to absorb, deflect, and reflect transmissions. While during this period, experiments with the heat ray were exploring how to transmit infrared content, these photographs cast bodies and objects as infrared receivers. More generally, these images underpinned a thermoceptive regime in which bodies and objects could become media for infrared transmissions and in which new practices of thermal manipulation could be developed.

By the 1950s and 1960s, thermography was used regularly in many fields. Zoologists captured thermographs of animals ranging from woodchucks to seals.[23] Infrared images were used to locate breast cancer, to reveal hidden layers of paintings, and to assist in the aerial detection of natural resources. They were even harnessed in emergent ecological studies of this period. Thermographs imaged the "thermal pollution" produced by electric power plants' cooling apparatuses as waste heat leaked into rivers and lakes.[24] The technologies of midcentury infrared imaging differed substantially from the early forms of infrared photography. As one study observes, "Modern infrared detection and image display systems bear little relationship to what in the past has been known as infrared photography."[25] Although, researchers argue, the average doctor might think about "infrared" in relation to surface veins, "the term infrared photography is a misnomer" because the earlier forms were capturing radiation only at the edge of visible wavelengths and bore "no direct relationship to the temperature of the object, and any heat recorded on them [was] the result of reflected infrared radiation, in contrast to directly radiated energy."[26] While Wood's images of trees and skies visualized short-wavelength infrared (just below the range of visible light), these new forms of infrared imaging were able to index a wider section of the spectrum, capturing longer and longer wavelengths.

Practically, these thermal infrared images recast the world not just as a field of bodies and environments that receive, relate, and react to radiation or as an evocative, sensory landscape but as an array of radiators. Thermal infrared cameras and light-sensing cameras both use a sensitive medium (film or digital sensors, for example) to index radiation. Light-based film typically captures light after it bounces off objects in the world. Only when it is pointed at an extraordinarily hot object—fire or a sun or molten lava—does it index objects as emitters of thermal radiation. Passive thermal imaging, on the other hand, captures emissions far below this thermal threshold and beyond perceptible levels—it offers a vision of the world not simply in terms of heat's effects but in terms of the heat actually generated by bodies and objects. It does so by gauging the infrared radiation emitted in the long-wavelength spectrum, which is directly related to temperature.[27] In this case, no visible light is needed to illuminate a scene. The scene itself is the illumination.[28]

In the 1970s, infrared cameras that could directly capture temperature were infused with a new imperative because of the energy crisis. Like so many thermal practices of this period, from the operation of thermostats to hue-heat research, the infrared image was enlisted as part of new social and technical approaches to energy conservation.[29] One of these was the energy audit—a detailed accounting of the energy consumption of individual elements in a building and buildings within a community. Although this mode of accounting first gained traction in the mid-1970s, it became an indispensable tool after the National Energy Act of 1978, which included the National Energy Conservation Policy Act. In these acts, the energy audit and varied forms of building insulation were highlighted as important techniques of energy conservation.

Discourses and practices of energy conservation seized on thermal imaging as a critical technology for the detection of heat loss. One report of that period argues that many buildings were constructed "in an era of cheap and abundant fuel, when insulation and thermal integrity were not primary considerations."[30] Thermography, especially new infrared scanners that made emissions instantly viewable on a black-and-white screen, was deployed as a solution: it visualized buildings to detect waste heat. Thermal aerial surveys were conducted, beginning with NASA's offices and extending across the United States, from South Dakota and Nebraska to schools in Illinois to the small city of Garland, Texas. As was true for many infrared images before, the analysis and use of these images cast thermal differences within a normative framework, designating normal and abnormal areas along the lines of temperature. Anomalous warm areas included smokestacks, skylights, and roof vents. Such features created what were termed "natural" warm points. Sites of "thermal expression" that were oddly shaped or excessively large were marked on the thermogram and slated for potential on-site investigation. Much of this was conducted using then-new digital processing techniques.

These images visualized heat in the tradition of medical and agricultural research imaging. In popular discourses they were situated in relation to these fields rather than in relation to military infrared. In one *Popular Sci-*

ence article, the author writes, "Technology has a new tool specifically designed to pinpoint . . . weak points in your home's insulation. It's done with thermography, the infrared imaging system that's been employed to spot cancer and other defects in the human body, and faults in machinery."[31] These infrared images were circulated much more broadly than prior thermograms in hospitals or scientific literature. High-contrast thermal divides differentiated unconstrained heat in ways that recalled the unconstrained veins, disease, and pollution of earlier images. These thermal images were a visual topography of dwindling resources, a map of their diffusion, and in the history of infrared photography, this was a key moment where a new thermoceptive regime enlisted the heat image as part of a digital feedback loop of environmental and thermal manipulation. "Irregular" thermal expressions directly motivated and justified the reinsulation of built structures: they helped to transform architectures' thermal mediation.

In 1977, even as thermal imaging sparked popular interest, such techniques were still not widespread, in part because of the high cost of instrumentation. Only a few companies offered commercial thermal cameras, including AGA, based in Sweden (where building thermography had been taken up much more rapidly than in the United States). AGA developed a thermal camera in 1963, and its models 680 and 750 were widely used for medical thermography before being used on buildings (figures 5.4 and 5.5).[32] But each of these scanners cost $36,000.[33] By the late 1970s, a movement had begun to expand the use and lessen the cost of thermal scanning as part of the "war on fuel waste."[34] In February 1977, Barnes Engineering Company released the ThermAtrace, which was easier to operate, did not require liquid nitrogen for cooling, and was significantly less expensive (only $6,000) than the AGA model. The following year, FLIR Systems was founded as a provider of thermal imagers for energy audits.

These energy audits and their infrared photographs, like Eastman Kodak's release of infrared plates, catalyzed an expansion in infrared imaging in the United States. Infrared was brought into people's homes and into new professions. These cameras capitalized on the potential of the thermal image not simply as a technique of identification or means of location but as an integral step in a feedback system of thermal control—particularly a system governed by economic efficiency and masked as environmental conservation. Heat images became critical media through which life and environments were thermally assessed and reengineered, and they recast the landscape itself as a site for thermal control and management—they helped to proliferate thermal media.

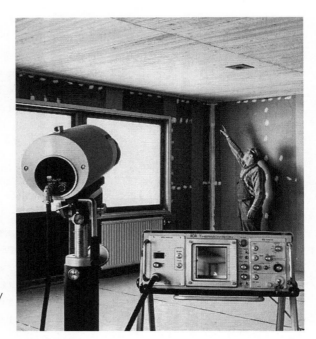

FIGURE 5.4. AGA Thermovision Model 680. From "AGA Thermovision," AGA Museum, accessed April 14, 2018, http://www.agamuseum.nl/page/thermovision.

FIGURE 5.5. AGA Thermovision Model 750. From "AGA Thermovision," AGA Museum, accessed April 14, 2018, http://www.agamuseum.nl/page/thermovision.

The thermal practices of the 1970s set the groundwork for the infrared imaging of the decades to follow. FLIR Systems became one of the leading developers of commercialized thermal cameras (and remains so to this day). AGA eventually segmented its infrared division as a subsidiary, AGEMA, which was acquired by FLIR in 1998. The Boston-based company Inframetrics, which developed the first TV-compatible infrared system in 1975, was also acquired by FLIR in the late 1990s. Today FLIR's cameras are critical tools not only for building contractors but also for a wide range of professional activities that involve managing the thermal landscape, from firefighting to farming. Thermal imaging is a standard practice in fields from veterinary medicine to the automotive industry. The use of heat to track disease, define anomalous bodies, and audit a landscape and the enfoldment of these uses into strategies of containment are residual in many of the commercial applications of infrared imaging.

Thermal infrared imaging is today a critical technique of environmental management. Like so many thermal technologies, it is enlisted not only in conservation efforts but also in in extraction. Infrared detection is often a first step in mineral and petroleum exploration. Thermal images, based on radiant data captured via satellite, enable the careful investigation of mineral deposits. This is possible because every element and compound has a particular thermal signature: quartz and hydrous silica can be detected by their emissions, around 8.40 μm and 8.95 μm, respectively. Thermal sensors can provide detailed data on soil types, vegetation, and other surface features at mine sites. With images such as the one of the Morenci mine in Arizona (figure 5.6), produced by the Satellite Imaging Corporation, companies can not only detect new deposits and expand their footprint but also increase the productivity of existing mines.

Thermal cameras are also deployed in agriculture. Today, thermal imaging is changing the way that farming works. Aside from the military, the USDA Foreign Agricultural Service is the largest user of satellite imagery in the United States government.[35] Drone and satellite images are used to monitor agricultural practice and production, to determine where droughts are occurring, and to ensure global food security. This is also occurring domestically. One journalist writes, "The first drones you're likely to see actually in use are . . . likely to be [close] to crop dusters."[36] A report by the Association for Unmanned Vehicle Systems International estimates that the largest potential market for drones will be in farming—that public safety is

FIGURE 5.6. Satellite thermal image of the Morenci mine in Arizona. From "Mining and Satellite Imagery," Satellite Imaging Corporation, accessed April 14, 2018, https://www.satimagingcorp.com/applications/energy/mining/.

only "a shadow market" compared to agriculture.[37] Equipped with thermal infrared imagers, these technologies track emissions of heat in the field, using the varying levels of heat released from different plant bodies to determine their health—or, in other words, their productivity, their potential as metabolic fuel for populations. This data is operationalized and integrated into circuits of networked control, determining how the thermal media of agriculture will be organized.

The etymology of the word *broadcast* is agricultural. In the mid-eighteenth century it meant "sown by scattering." Seeds were not planted one by one but scattered across a field. One to many. The term was adopted to describe radio, television, and electronic media only in the twentieth century. Alongside the larger transition from broadcast to narrowcast and microcast media, thermal imaging in farming is moving toward a microcast model termed "precision agriculture." If before, farmers broadcast seeds, water, fertilizers, and other elemental media to stimulate growth, today they are increasingly targeting narrow areas, particular sections of the field, for their distributions.

Thermal sensing technologies are not simply looking for heat. In the images they produce, heat is an indicator of other processes. As plants grow, they absorb sunlight, either using it for photosynthesis or converting it into heat. Tiny pores on plants' leaves allow heat to be released, a cooling-off mechanism not unlike those of the human body. But when a plant doesn't have enough water or is under stress, the amount of heat it emits changes. By reading the varying emissions of plants, alongside the emissions of the soil, these technologies enable farmers to determine what sections of their field need mediation and to deploy agricultural technologies more efficiently. These minute thermal differences are mobilized productively to increase margins. As is true in the recent development of digital thermostats and personal technologies of temperature control, thermal difference is the key variable to be located and assessed as a means of generating profit.

Farmers, like physicians and building contractors, have long used their own bodies to assess their environments, cataloging signs of heat via the senses of both sight and thermoception in order to determine where and how to mediate that heat. But what was measured through eyes and skin is now registered by a digitized sensor, and machinic thermal vision is automatically incorporated into systems of thermomanipulation. One physical scientist reflects, "In the early days, when farmers had small fields, they knew from practical experience which sub-areas were wetter and more fertile," but now, with land of over "50,000 acres, farmers start to lose touch with their fields."[38] In large-scale agriculture, drone imaging, and global food security, the body is no longer a primary capture device for heat's transmissions—and this allows a dramatic shift in the scale of heat manipulation.

Like the many previous practices of infrared imaging, the infrared targeting of agricultural landscapes defines an entire field as a set of thermal zones, healthy and unhealthy, in need of pesticides or water, or sites to be ignored. It is linked to increases in other scalar technologies—drone dusters, self-driving tractors, and robotic processing technologies—all of which generate and depend on massive amounts of data. And like so many other digital thermocultures, agricultural narrowcasting or microcasting is being sold as an ecological endeavor. If before, pesticides were broadcast across a field, heat sensing and precision agriculture enable their application only where deemed necessary. Water is moved through irrigation channels only when necessary. It is directed to where the thermal image tells the farmers to distribute it: micromanagement farming. Some imagine that this will dovetail with the organic food movement and that customers will be willing to pay for drone-guided agriculture.[39] But rather than a radical break from the history of commercial farming, Christopher Miles shows, "the truths that digital sensors and algorithmic processing speak are the expression of a normative function: the rational logic of capitalist production."[40] Precision agriculture's thermal images do not simply display temperature differences but express thermal normativity and extend the historical use of infrared to maximize production.

In wildlife management, thermal infrared images are also being enfolded in logics of ecological narrowcasting and landscape micromanagement. Conservation biologists describe the utility of infrared cameras in a variety of situations—from detecting polar bear dens to counting bighorn sheep to tracking and hunting feral goats on islands.[41] In deer-management programs, infrared cameras register the number of deer in an area and subsequently help define parameters in which sharpshooters can kill them. Such imagery is also critical for wildlife media that use the camera to extend the human gaze further into the "natural" world. In one television show, *Extinct or Alive*, the host describes how a FLIR infrared camera is critical to efforts to find the Tasmanian tiger.[42] Animals are made visible on the basis of their thermal signatures, either indexed as a part of a population or singled out as spectacle.

Thermal imaging has also been widely used for conservation in the enforcement and containment of designated wildlife zones. In 2012, the World Wildlife Fund (WWF) was awarded a $5 million Global Impact Award from Google to "harness technological innovation to stop conservation crime."[43] The Wildlife Crime Technology Project, which emerged out of this grant, aimed to stop the "epidemic" of global wildlife poaching, with sites in Kenya, Malawi, Namibia, South Africa, Zambia, and Zimbabwe. While the project initially worked to develop cell phone–based patrol networks, drone surveillance, and wildlife tracking tags, each of these technologies faced limitations: the shortage of cell service, the difficulty of securing a power supply, and the banning of drones. In the project's second phase, the WWF began to test thermal cameras as a means of monitoring wildlife parks at night. In 2016, they installed FLIR imagers in Kenya, one coupled with artificial intelligence in Lake Nakuru National Park and another atop a mobile ranger at Maasai Mara National Reserve. A publicity video boasts that "rangers will own the night" and features images of the mobile FLIR camera being monitored by rangers (figure 5.7).[44] The following year, the WWF collaborated with digital hardware company Cisco Systems on a project at Kafue National Park in Zambia that included FLIR cameras. They also installed a

FIGURE 5.7. An image displayed prominently on the World Wildlife Fund's Wildlife Technology Crime Project web page features a FLIR camera as part of an antipoaching campaign. From "Wildlife Crime Technology Project," *World Wildlife Fund,* accessed April 5, 2021, https://www.worldwildlife.org/projects/wildlife-crime-technology-project.

long-range FLIR detection system in Malawi. The WWF publicly announced that many arrests were made thanks to thermal technology.[45] Its digital media partnerships, especially with FLIR in developing the mobile Kenyan unit, made international news and received extended coverage in publications such as *Wired, Medium,* and *National Geographic.*[46]

In the Wildlife Crime Technology Project, FLIR cameras were enlisted as part of a digital operation to render the landscape as a thermal gradient that could be parsed by both artificial and human recognition. While a long tradition of conservation biology, especially following the development of infrared motion detection in the 1980s, has used infrared cameras and camera traps to capture target animals, the WWF uses these to recognize humans and to capture poachers.[47] This is a prime example of what Elizabeth Lunstrum calls "green militarization," the use of military personnel, technologies, and training as part of conservation efforts.[48] "Much of what conservationists want to do already exists, but is in the hands of the military," says Eric Dinerstein, the vice president for conservation science at the WWF.[49] In its "democratization" of thermal imaging from energy audits onward, FLIR is a company that makes such militarized thermal visions more affordable and accessible. At the same time, antipoaching cameras represent a convergence of the military history of thermal imaging, which uses body heat to locate human targets, and medical and environmental imaging, which segments the landscape into healthy and unhealthy zones, with cameras reading for invasive cancers and pests. In this orientation, the cameras are calibrated so that to show up on the screen, as heat, is already a crime, an invasion, and an infection.

In recent years, thermal imaging has spread across sub-Saharan Africa. Drones with thermal cameras fly over private game reserves. Foundations donate money and equipment, including aircraft with infrared detectors, for aerial surveys. Antipoaching intelligence organizations deploy FLIR cameras for park observation. These efforts are part of an expansive security industry that links game reserves and parks into a broader network of thermal vision. SecuSystems, a Johannesburg-based company, installs and distributes FLIR cameras not only for antipoaching activities but also for mine surveillance. TeleEye South Africa, another security company, unveils a line of FLIR cameras that it describes as "ideally suited" to combatting rhino poaching.[50] Cape Town–based Timeless Technologies offers thermal imaging systems for conservation as well as for surveillance by casinos, law enforcement, and cities. Even when they are not connected to stopping "wildlife crime," these projects are sold as environmental projects because they are deployed to protect renewable-energy plants and feature installations powered by "self-sustaining green energy."[51] The WWF's thermal visions are technologically, industrially, and socially enmeshed with an array of other forms of containment and surveillance.

This use of thermal imaging, like precision agriculture, is described by its advocates as a win-win solution: it saves animals and rangers, promotes tourism, stops crime, and promotes digital innovation. It draws funding and support from technology companies, militaries, and conservationists. But this thermal vision is a biopolitical form that, like biodiversity conservation itself, defines some forms of life as valued (such as elephants, miners, and estate residents) more than others, such as the poachers hunting for bushmeat in Kenya, crossing the border from Tanzania, or the Black hunters in South Africa who are regularly hunted and murdered by antipoaching forces at the same time that white American game hunters pay massive sums to kill game animals. In these projects, and especially in South Africa, where thermal imaging is deeply embedded in apartheid infrastructures, thermal vision connects racialized securitization to global technological industries and masks it as ecological and economic conservation.

The representation of poachers' bodies as heat, and heat in bodies, is inflected not only by the tradition of military-style *Predator* vision but by a set of perceptual modes that can be traced back through medical imaging and energy audits, which ground thermal vision in structures of thermal normativity. In such thermal visions, to be outside a thermal zone is to already be diseased, abnormal, and in need of correction. In the thermal drone's vision above the South African reserve, humans become targets and the animal's body is enfolded into the mass of safari images: bundles of glowing heat. Situated alongside such images is a long history of racist thermal knowledge that depicts Blackness as always already too hot. Far from the thermal vision's claim to neutrality, these are deeply racialized images that project a white and Western technological imaginary as part of a strategy of containment, management, exclusion, and violence against entities based on their body heat.

AGAINST THERMAL NEUTRALITY

In Richard Mosse's exhibition of thermal images, titled *Heat Maps*, one panoramic black-and-white photograph reveals a stadium in the mountains (figure 5.8). The bleachers are empty, the sky a blurred gray. On the playing field, a grid of tents houses unseen bodies. People are visible here and there in the landscape. Their bodies are small and completely white, blown out— so white their features are not visible. As in many thermal infrared images, whiteness here signifies marked bodies. In this case, these are refugees who have been targeted, housed in camps, and made visibly alien in their surroundings by Richard Mosse's thermal infrared camera.

Even in this detailed and densely populated thermal landscape, it is impossible to glean information about who these people are or where they are located. They are, in the exhibit's description, isolated, disembodied, and reduced to a "mere biological trace."[52] To shoot the images in *Heat Maps*, Mosse used a thermal imaging camera that was designed for military applications, including battlefield awareness and target tracking, as well as border control. In international law, the camera itself is classified as a weapon. As the gallery that exhibited the photographs describes it, the camera "dehumanizes its subject" and recasts them as "bare life."[53] Like so many military, policing, and carceral uses of infrared imagery, vision itself is here a form of thermal violence, a means of reducing a subject to their thermal emissions, of turning life into heat that can be used in forms of biopolitical management.

In the book that accompanies this exhibition, *Incoming*, and the fifty-two-minute multiscreen video of the same name, the aesthetics are inverted

FIGURE 5.8. This 2017 photograph from Richard Mosse's exhibit *Heat Maps* shows Hellinikon Olympic Arena in Athens. From Richard Mosse, "Hellinikon Olympic Arena," from the series *Heat Maps*, 2016. Digital C-print on metallic paper.

but the depersonalization of refugees remains just as powerful. The book features images of a clearly bounded sun. Military aircraft glow bright white against a dark sky, as do missiles and the debris of explosions. Refugees huddle on boats and behind fences, their skin luminous. Mosse captures close-ups and extreme close-ups of their bodies, faces, and hands, sometimes from miles away. The Horizon Medium Wave Infra-Red camera is capable of detecting a human body at over thirty kilometers and identifying it as such from six kilometers.[54] The images of these people reveal intimate moments, yet the camera often flattens skin texture and makes their eyes bright white, blank and radiant against the darkness of their bodies. Many of the images feature a stark crispness and fine detail that elicit a sense of militarized geolocation, as well as indefinite, blurred traces of motion that evoke a sense of haptic visuality.[55] Mosse's shots in the book and video do to location what *Heat Maps* does to bodies: they detach intimacy, violence, and death from their specific environment, making any location interchangeable with others. The images are often framed so that the edges of the photograph cut through the phenomena under view, severing them from context. Mosse states that this is a key limitation of the camera itself: "It is unable to capture a wide-angle establishing shot, incapable of 'setting the scene.'"[56] But even when there are opportunities to provide a scene, the photographer refuses to do so. Rather than "rescue this apparatus from its sinister purpose," Mosse argues that they tried to "work the technology against it-self."[57] The world is defamiliarized and mortality itself is foregrounded. Just as Mosse can sit miles away and watch intimate transactions, the viewer can likewise feel these images, and the suffering of bodies, while holding them at a safe distance.

There is no critical distinction in these images between refugees and sol-diers, between military ships and traffickers' boats, between safe havens and targets, or between weapons and bodies: all are glowing and set against a dark background. One night, Mosse recounts, through the lens of their camera the team witnessed a boat of about three hundred refugees sink. Images captured the heat prints of hands on cold bodies, visually indicating the transmission of life as the transmission of heat. Such images resonate with prior infrared art, including Terike Haapoja's installation *Community* (2007), which documents the cooling of animal bodies after death. They also powerfully echo the descriptions of drone operators. Brandon Bryant, for example, remembers watching a person roll around on the ground: "I watched his life's blood spurt out in the rhythm of his heart. It was January in the mountains of Afghanistan. The blood cooled. He stopped moving, eventually losing enough body heat to become indistinguishable from the ground on which he died."[58] For Mosse's camera and for the drone operator, the most important difference is that between heat or nonheat, which is often that between life and nonlife, between target and backdrop.

The descriptions of these images, by Mosse and others, try to buttress in-frared's claim to neutrality: technologies of heat detection and representa-tion are described as indifferent to differences between bodies and, as such, to race and gender. Because the only significant difference is that between life and nonlife, many claim, there cannot be an inherent racial politics to this apparatus or to this form of thermal vision. Mosse argues, "The camera is colour-blind—registering only the contours of relative heat difference."[59] Thermoemissive bodies, registered from above or afar, communicate only the distinct feature of aliveness. But what Mosse's camera does, Niall Mar-tin points out, is reveal "a migrant body in which earlier signs of blackness have been detached from race and reconfigured purely as heat."[60] In a de-scription of the drone's "hunt for heat," Lisa Parks argues that these kinds of visual surveillance practices extend "beyond epidermalization," producing what she terms "*spectral suspects*—visualizations of temperature data that take on the biophysical contours of a human body."[61] Far from eliminating ethnic and racial differentiation, Parks shows, these practices restructure difference "along a vertical axis of power and [recodify it] according to issues such as moving to or being in certain places at certain times, being in the vicinity of other suspects, driving certain vehicles, or carrying certain objects with certain temperatures or shapes or sizes."[62] Difference is integral to thermal vision and in the forms of thermal violence it enacts. Moreover, it rarely exists simply as a mere distinction between heat and cold. As the cases throughout this chapter reveal, it materializes and makes sense only within a broader structure of thermopower.

Collectively, these uses of thermal infrared imagery reveal how temperature is manipulated not only through thermal infrastructures, environments, architectures, food, movement, and metabolism but also through vision and visual technologies. These images transduce the heat of bodies into an image that is integrated into feedback loops of targeting, surveillance, and the consumption of suffering. Crystallized within thermocultural practices, the infrared camera is not objective—not simply because it is used in practices of targeting but because it divides the landscape into regular and irregular zones, structures it in relation to thermal normativity, and casts it in relation to social and cultural values. In medical practice heat often indicates disease, in environmental management it often indicates pollution, and in the restructuring of the 1970s home, it indicates leakage and waste. In these heat maps, what becomes visible is abnormal: to vary in temperature is to invite environmental manipulation and modulation, often through strategies of insulation and containment.

Looking at the history of these images and at other forms of nonmilitary infrared photography—even those that are not "thermal" imaging per se, such as Robert Williams Wood's trees—reveals that *Predator* vision, so often described as indifferent to race and gender, often carries with it assumptions about bodily normativity that are inflected by the longer history of commercial infrared photography. The thermal target is not simply a spectral subject, a thermal trace of bare life; it is abnormal in its radiance. The calibration of the camera, which renders the subject in high contrast to the background, and the hue-oriented thermal vision that naturalizes heat as red, itself an aesthetic choice, marks bodies themselves as *too hot*, a visual correlate to the disease, the cancer, the escaped heat. Digital processing technologies find optimal data in the infrared image, the high contrast of which aids in object edge detection. Reading *Predator* vision as one part of a broader set of thermal visions makes clear that the practices of infrared, even as they bring viewers into a world where they "do not see things as they are," anchor these apparently objective and nonhuman landscapes in normative stratifications.[63]

Although the thermal image is widely used for security purposes, in everything from policing to pandemic management, these are not generally produced by military thermal cameras. The industry of infrared imaging extends far beyond the military, and has since the early twentieth century. Agriculture, architecture, construction, medicine, and conservation, among other fields, capitalize on the heat image. These commercial uses of infrared cameras, which are often paired with the transformation of the landscape into a thermal medium, are a substantial source of income for camera manufacturers, including FLIR Systems. As a result, the historical and contemporary capacity of thermal infrared in everyday environments underwrites the affordability of infrared cameras for policing and strengthens their use as a tool of targeting more broadly. Thermal vision has helped mold the landscape in relation to a radiographic episteme, justifying numerous other forms of manipulation.

THE POLITICAL POTENTIAL OF THERMAL VISION

Seeing by heat takes on a new significance in light of both social and thermal volatility. In fields from agriculture to conservation, infrared images are now viewed as a key tool of ecological management and as a possible means of mitigating climate change's effects. The environmental potential of thermal vision has also been harnessed in other representations of temperature, especially in the use of color and design. In a special issue of *Building Research and Information*, "Counting the Costs of Comfort," Sue Roaf, Luisa Brotas, and Fergus Nicol argue that the costs of thermal comfort must now be calculated in social and environmental, not just economic, terms. In a description resonant with an earlier era of hue-heat research in the 1960s and 1970s, they call upon building designers to think about comfort expansively and to reflect on the continuing controversy in thermal comfort studies over the "mismatch between the accepted definition of thermal comfort as 'the state of mind which expresses satisfaction with the thermal environment' and the methods used to investigate comfort which use physics and physiology but spend little time on the way individuals actually perceive comfort."[64] In short, they call for techniques that change the perception of temperature without actually lowering or raising the temperature, for a haptic visuality that can affect the body through sight. If thermoception can be synesthetically engaged through vision, then a sense of climate might be managed through a set of visual practices.

Just as the 1970s energy crisis shaped infrared imaging, climate change is reactivating thermal vision with social and political potential. A wealth of research, industrial design, and corporate practices seek to manipulate the perception of temperature, especially through the use of color—if not to lessen the dependence on fossil fuel–based thermal manipulation like air-conditioning, at the very least to reduce their own energy costs. As just one example, scientists at the German Aerospace Center published a study on the effect of colored light—blue, yellow, green, and violet—on the perception of temperature. They solicited almost two hundred participants to sit inside a mock aircraft cabin, interspersed with "thermo-dummies" that measured the actual temperature, and then altered both the actual temperature and the color of LED lights. What they found validated the hue-heat hypothesis: subjects felt warmer, and more satisfied, with yellow lights. They advised the airline industry that yellow "contributes, directly or indirectly by means of a kind of 'psychological' warming-up, to the overall satisfaction with the climate."[65] Their findings ran counter to existing trends in cabin lighting, which generally uses blues and violets to instill a sense of calm and soothe passengers. Based on these results, the researchers suggested that deploying color to alter perceived temperature might lead to substantial ecological benefits. Like the infrared camera, these design practices recast the environment as a medium for thermal communication: alongside personal thermostats, fans, and wearable air conditioners, ambient imagery is a means of cooling and warming climate-impacted subjects.

Beyond its immediate felt, tactile, and embodied effects, thermal visions— whether the colors of interior design or the heat images of infrared cameras— have the political capacity to naturalize thermoceptive regimes. Just like the smart thermostat described in chapter 1, thermal images are today offering a vision of the world as a radiant force, but one that should be managed through digital processes. At the same time—and as is true for many contemporary thermal media—the infrared camera also claims to offer a more objective view of temperature. And it is a key means by which temperature, used as a metric for all bodies in all spaces, becomes the new visual language of objectivity. Yet in reality, as these cameras are used increasingly to take the temperature of both bodies and their environments, they bring with them a set of invisible histories of normativity, racialized and gendered thermal distinctions, and assumptions about ecological and bodily management. In other words, they are vectors for the spread of thermopower.

6 COMPUTER THE COLDWARD COURSE OF MEDIA

The coldest known place in the universe, where the temperature is −273.15°C/
−459.67°F (only slightly above absolute zero), is not in outer space or in the
depths of the ocean. It is in Yorktown, New York. In an IBM research facility,
the inside of a translucent glass box is cooled to 15 millikelvin. These same
superlow temperatures are manufactured by D-Wave in Burnaby, British
Columbia, where a ten-foot-by-ten-foot black box is also cooled to around
−273°C/−459°F. This extraordinary cold, maintained at great cost and care
using cryogenic refrigerators and liquid helium, has the sole purpose of sta-
bilizing qubits—the bits of quantum computers.

Most computers are digital. They rely on binary transistors that switch
a current on or off. These states are then interpreted as the zeros and ones
that constitute digital information. Transistors form the material basis for
modular logics, which as Tara McPherson demonstrates, scaffold a broader
cultural operating system in which the world is perceived as isolated bits.[1]
In quantum computing, hailed as the first true paradigm shift in computing
since the adoption of the von Neumann architecture after World War II, the

qubit—the quantum bit—is not fixed as either a zero or a one. It can be represented as a zero, one, a superpositioned zero and one, or values between zero and one. As Derek Noon describes in an ethnography of quantum computing, while in the "file clerk approach" of classical computers, each bit is given a fixed value that is independent, stable, and stored, qubits are entangled with one another: their value can be influenced by other qubits. Inside the quantum computer, the state of the qubit is undetermined until it is put through an energy program that transitions it into a "static value."[2]

As they mobilize quantum mechanics as part of their operations, quantum computers are inflected by any intra-actions with their observers and the physical world. As a result, the companies that manufacture them take extreme steps to separate the qubits from their surrounding environment. First among their concerns is the reduction of thermal entanglement, also known as "thermal noise."[3] Instead of silicon, the base material of quantum computers is niobium, used in part because of the way it reacts to temperature. Near absolute zero it exhibits quantum effects that can be "isolated and controlled" as qubits.[4] But that means quantum computing performance is directly linked to temperature: functionality depends on remaining below 80 millikelvin (−273.07°C/−459.53°F) and systems operate close to 15 millikelvin (−273.14°C/−459.64°F).[5] An infographic from IBM, "Inside Look: Quantum Computer," focuses not on a quantum computer but on the dilution refrigerator that houses it (figure 6.1). The infographic highlights some of its two thousand components using bright-blue text boxes: cryogenic isolators, a cryoperm shield, a mixing chamber that supplies cooling power, and the application of attenuation to "protect against thermal noise."[6] Quantum computing, IBM tells the reader, is a "march to absolute zero."[7] The precondition for quantum computing, and the only environment in which entanglement can be harnessed as computational force, is the extreme cold.

Heat has long been a by-product of traditional digital computing, an exhaust produced by the interactions of its material components, a kind of friction in the material world. In turn, as Finn Brunton characterizes it, "the work of computation is the work of managing heat."[8] Without expansive cooling infrastructures to offset the massive amounts of heat generated by digital systems—and to reduce thermal entanglement—information would be incorrectly registered or not registered at all. Hard drives would burn out and processors would overheat. Data centers would cease to function and internet traffic would come to a halt.

Inspired by the unique temperature dependency of digital systems, artists, authors, and scholars have drawn attention to the work of cooling.

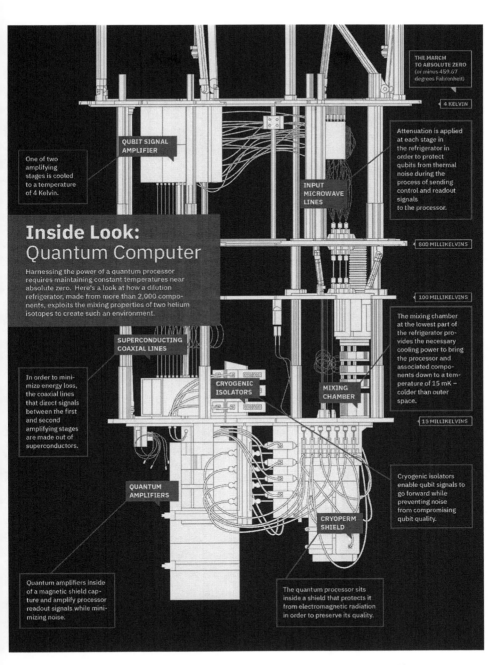

The following text appears as labels within the infographic:

THE MARCH TO ABSOLUTE ZERO (or minus 459.67 degrees Fahrenheit)

4 KELVIN

QUBIT SIGNAL AMPLIFIER

One of two amplifying stages is cooled to a temperature of 4 Kelvin.

INPUT MICROWAVE LINES

Attenuation is applied at each stage in the refrigerator in order to protect qubits from thermal noise during the process of sending control and readout signals to the processor.

Inside Look:
Quantum Computer

Harnessing the power of a quantum processor requires maintaining constant temperatures near absolute zero. Here's a look at how a dilution refrigerator, made from more than 2,000 components, exploits the mixing properties of two helium isotopes to create such an environment.

800 MILLIKELVINS

100 MILLIKELVINS

SUPERCONDUCTING COAXIAL LINES

In order to minimize energy loss, the coaxial lines that direct signals between the first and second amplifying stages are made out of superconductors.

CRYOGENIC ISOLATORS

MIXING CHAMBER

The mixing chamber at the lowest part of the refrigerator provides the necessary cooling power to bring the processor and associated components down to a temperature of 15 mK – colder than outer space.

15 MILLIKELVINS

QUANTUM AMPLIFIERS

CRYOPERM SHIELD

Cryogenic isolators enable qubit signals to go forward while preventing noise from compromising qubit quality.

Quantum amplifiers inside of a magnetic shield capture and amplify processor readout signals while minimizing noise.

The quantum processor sits inside a shield that protects it from electromagnetic radiation in order to preserve its quality.

FIGURE 6.1. An IBM infographic on quantum computing focuses on the dilution refrigerator. From "Quantum Computing at IBM," IBM, accessed February 1, 2019, https://www.research.ibm.com/ibm-q/learn/what-is-ibm-q.

They have created experimental videos that feature air conditioners. They have documented how companies recycle waste heat and transform it into a thermal product.[9] They have critiqued the extraordinary economic and environmental costs of cooling systems (maintaining the cold constitutes the bulk of energy expenditure of contemporary computing, whether digital or quantum). And they have tracked the efforts of infrastructure builders to maintain data in warmer temperatures. This art and research, both scholarly and popular, constitutes another kind of thermal media, a discourse that reveals how digital systems are tethered to an underlying thermal regime.

This chapter documents how media came to be so integrally linked to thermal regimes. Managing heat through artificial cold is an integral part of media history. In turn, media's need for thermal manipulation has been a motivation for developing new forms of cooling—the quantum computer and its novel refrigeration system are only the latest in this march. The first section of this chapter dives into the foundational thermocultural practices of early twentieth-century media. The first air conditioner, as described in this section, was initially created not for people but for media. As it was installed in the print industry, temperature control made color printing possible. Following this, air-conditioning accelerated the laboratory processing and theater projection of film, stabilized phonograph production, and was used in new television studios, among many other media architectures. In each of these cases, thermal contexts were manipulated in order to stabilize media into a standard object, ensure its reproducibility, and scale up production. In short, thermal technologies, including air conditioners, architectures, and thermostats, have always prevented the entanglement of modern media with its thermal contexts. They are what make modern media into mass media.

This history reveals how thermopower is enacted in the calibration of media themselves: thermal technologies compose an infrastructure for all modern media objects. And yet, even as industry-deployed thermal technologies stabilize media's materiality during manufacturing, after media leave the warehouse they inevitably remain sensitive to temperature. It is not only climate but people's bodies that threaten media: as Fernando Domínguez Rubio argues, the heat people generate destabilizes the conditions for media objects to even exist, especially in the museums where they are preserved.[10] The second section of this chapter turns to the thermal problems of media consumption and use, from the warping of phonograph cylinders to the explosion of nitrate film. The threat of thermal entanglement, which is so acute in the case of the quantum computer, has always been a problem for media. Too much heat, too much cold, or simply too much thermal change rapidly degrades communications technologies.

In order to stabilize the media object over time, a second form of thermal management is enacted. Cultural theorists have recently begun to investigate cryopolitics, the political operations of all kinds of artificially cooled environments, in which "the regulation of life is extended indefinitely through technoscientific means such that death appears perpetually deferred."[11] Cryopolitical technologies freeze media objects in particular shapes: they are a form of arresting, of halting a crisis or movement, such that media remain a stable object that withstands time.[12] Preservation through cooling is a thermocultural practice, a cryogenic culture, as Joanna Radin and Emma Kowal explain, infused with the impulse to "cheat death" by suspending, slowing down, and freezing.[13] Refrigeration and artificial cooling are as materially significant to the traditional media landscape as they have been to food systems and scientific practice.[14] Describing the work of these cooling technologies in ongoing media operations, this section reveals how the cryopolitics of media operates as a fundamentally conservative thermal regime with the aim of keep media's forms intact.

As media manufacturing and use is tied to thermal maintenance, media technologies are embedded in a broader thermal regime. Media technologies and practices are optimized for some climates and not others. As a result, choices about media operation privilege particular environments and reinforce geopolitical relationships. For most of the twentieth century, the dependence on cooling and climate stability, especially on environments with low humidity, ensured media continuity and historical preservation in locations with extensive atmospheric control (and the institutional and energy systems to support it). Returning to digital systems in the chapter's final section, I show that as more and more thermal maintenance and cooling systems have been required, this has resulted in a coldward course of media. Here I intentionally recall S. Colum GilFillan's racist and environmental determinist argument in "The Coldward Course of Progress." In this 1920 text, GilFillan argues that as mankind expanded its thermal range to cooler climates, people had to develop clothes, architecture, and instruments of heat production, all of which ultimately resulted in a greater capacity for civilization.[15] One diagram in GilFillan's article purports to chart the relationship between annual mean temperature and world leadership (figure 6.2). This image, resonating with centuries-old arguments by environmental determinists, naturalizes white supremacy and Western domination using thermal terms. Supremacy for these authors emerged from a natural mastery

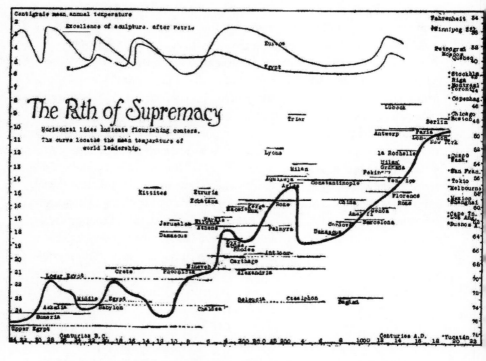

FIGURE 6.2. "The Path of Supremacy" purports to chart a connection between mean annual temperature and the "flourishing centers" of world leadership. From S. C. GilFillan, "The Coldward Course of Progress," *Political Science Quarterly* 35, no. 3 (1920): 395.

of the cold or the natural effects of colder climates—and as described in prior chapters, such thermal knowledge was used to support slavery and colonization.

In this chapter, I show how the coldward course of media over the twentieth and twenty-first centuries, like GilFillan's "coldward course of progress," is neither a natural teleology nor a simple extension of "natural" environments through media production. The coldward course of media is a social project that explicitly connects the "progress" of communication to temperature control and, as a result, ties it to the social, economic, and political conditions for this control. It is a project of designing media for particular environmental conditions, ensuring homogeneity across media objects, and maintaining the consistency of media. The coldward course of media is predicated on a regime of thermopower that invisibly and systematically re-embeds media production, distribution, and access into the control of institutions and actors with the ability to engineer stable climates.

A new geopolitics appears to emerge from this need for cooling: a cold-ward course of computing that privileges northern geographies over tropical ones and further empowers companies to exploit the environment's thermal affordances. The final section of this chapter describes the ways that the social project of media cooling not only is shaped by the impulse toward sameness, toward alignment, and through the repression of difference; it also privileges cold locations, especially in northern countries, for the architecture of the internet. In turn, the internet is being built with a specific natural geography and climate in mind. Thermal environments, natural or artificial, are increasingly a site for media's exploitation: as Julia Velkova has shown, Nordic countries in particular are becoming a new zone of extraction because of their thermal affordances.[16] And yet even while some northern environments have a climatic advantage, climate alone does not determine where the internet is built: those with the capacity for cooling, cheap power, and advanced thermal technologies will have more access to and control over contemporary media, especially in a time of increased thermal variability.

THE TEMPERATURE PROBLEMS OF MODERN MEDIA

Media's coldward course began in the print industry in the late nineteenth and early twentieth centuries. During this period, printers and lithographers were grappling with the problem of fluctuating temperatures and humidity. As one printer notes, "There is probably no other craft that is so dependent on the preservation of an even temperature as presswork."[17] Difficulties extended from heat's "chemical action" on ink to the "generation of electricity in paper" during cold weather.[18] Paper was a fickle material, expanding and contracting depending on climatic conditions. One critical obstacle, especially in the development of color printing, was improper registration: if temperature and humidity shifted as individual sheets were passed multiple times through a printer, the inks would be overlaid imperfectly. The image would vary from the intended production and each image would be slightly different from the others. Another writer observes that the susceptibility of paper to climate is "one of the greatest difficulties the color printer is up against."[19] Thermal entanglement was preventing both consistency and standardization.

Printers had temperature problems not only with the external climate but also with the indoor climate, especially the excess heat generated by the presses. Some pressrooms were full of hot air, and in these conditions, the pressfeeder was tasked with inserting page after page into the machine in the exact same position. The editor of the *Inland Printer* writes, "No occupation in the printing trade is quite so monotonous as that of the pressfeeder, yet on none is greater accuracy entailed." It required "sustained skill and steadiness of nerve," particularly given the "enervating conditions."[20] Thermal entanglement extended to the workers' bodily performance: heat affected their ability to accurately feed paper into the machine.

The problem of uneven temperature materialized in both heat and cold, in the paper and in the bodies of pressfeeders, and in the machinery itself. Even in the early twentieth century, it was still common practice in many pressrooms to turn off the heat at the end of a winter day, leaving the iron and steel machines to cool overnight. The editor of the *Morning Citizen* reported that every morning, the machines were cold: "It requires considerable time to warm up. . . . if one put his hand upon a press or cutter he experiences a cold shock," and it was not until midday that everything worked "just right."[21] The expansion and contraction of metals caused problems over time, especially in composition rollers, which were "more susceptible to temperature."[22] Although the thermal sensitivity of organic substances such as paper and bodies was often observed, metals were also thermally entangled.

In the absence of technologies for precise temperature control, achieving consistency across texts and images required a great deal of time and money. Allan Turck lamented the amount of money lost each year on postage stamps that would expand or contract before they were cut and reported that "the United States Government has a standing offer of a large reward for the man who invents a printing-paper that will not distort under adverse conditions."[23] Some printers waited until the climate was suitable to run their presses, but slowing or shutting down compromised the quick timetable demanded by the circulation of news. Presses in different climatic zones thus operated at different speeds and with differing degrees of accuracy. A British lithographer observed, "German lithographers are greatly helped by the differences in their climate. With our varying temperature and our rather humid atmosphere, it is almost hopeless to register with precision sheets of fifty inches by forty inches. Yet Germany does this easily."[24] Temperature shaped the speed, cost, and potential formats of the printing industry, and as a result, the production of print media reflected thermal variations across days, seasons, and climates.

Attempts were made to address this problem by manufacturing paper with a lower sensitivity to temperature and humidity, but ultimately the solution adopted was environmental control—and particularly the convective medium of air-conditioning. According to its well-rehearsed origin story, the father of air-conditioning, Willis Carrier, was invited to the Brooklyn printing company Sackett and Wilhelms in 1902. At the time, the company was dealing with the typical problems of color printing: as ink was applied one layer at a time, images were misaligned when the paper stock expanded and contracted, causing delays in the production schedule, producing a massive amount of waste, and resulting in poor quality.[25] Carrier designed an air-conditioning system to regulate temperature and humidity in the plant, improving the reproduction of color images in *Judge* magazine, one of Sackett and Wilhelms's most important clients. As it was deployed across printing plants, air-conditioning facilitated the standardization and precise replication of paper-based media. In turn, it linked the expansion of color printing to a regime of thermopower, tethering the standardized image to an infrastructure of temperature control.[26]

Following the success of the printing industry, this regime of temperature control came to stabilize many other forms of modern media in the early twentieth century. In early film manufacturing, variable temperature and humidity were affecting the printing of film. At the Celluloid Company's plant in Newark, New Jersey, humid weather caused white specks to form on the film, which translated to white spots on the screen when the film was projected.[27] While damp air could alter the image, dry air could make the film brittle. And if the air was not clean, the gelatin coating of emulsioned film stock would behave "as a sticky paper towards dust particles in the air."[28] "One of the most important factors [in the film laboratory] is the problem of removing dust," engineers argued.[29] Incorrect temperatures themselves could lead to "curling and distortion" of the film's gelatin surface, and "in the swollen or jelly stage (after washing) the film may even be melted by too warm an atmosphere."[30] Heat exposure also heightened film's sensitivity to abrasion, and when running film between leathers, "scratch marks [would] be very easily produced in a too-warm laboratory."[31] As in the case of the overheated pressfeeders, too-warm temperatures could also warm the bodies of the workers, whose "perspiring fingers" might leave sweat drops on the film prints.[32] These environmental forces—temperature, humidity, and particles—together formed a significant threat to the standardization of the cinematic image.

As in the case of color printing, air-conditioning systems promised to stabilize the environment and, as a result, ensure the precise replication of the cinematic image. After the successful installation of air-conditioning in Sackett and Wilhelms, Carrier installed an air-conditioning system to eliminate the white specks that formed on film at the Celluloid Company. Atmosphere- and temperature-control systems were deployed across new film manufacturing facilities throughout the 1910s. In 1914 the Willat Film Manufacturing Corporation erected one of the largest film plants in the United States in Fort Lee, New Jersey. Its "engineered film laboratory" featured an elaborate system of temperature and humidity control, including a drying room with conditions controlled with ice and an "air washing" apparatus installed by the Carrier Corporation.[33] The Universal Film Manufacturing Company broke ground on a half-million-dollar facility in Fort Lee the following year, which also included an air-conditioning apparatus that kept the air in the drying room at 24°C/75°F.[34] In 1921 *Scientific American* reported, "Mechanical weather is used in nearly every branch of the laboratory work of the motion picture industry from the drying of the celluloid film base to the drying of the finished film."[35] In the 1910s, Fort Lee was a hub of advanced techniques in atmospheric control.

In these facilities, air-conditioning maintained optimal humidity, kept the film from becoming dry and brittle (and from breaking), and contributed "to the production of a soft, pliable, flat film."[36] C. A. Willat, head of the Willat Film Manufacturing Corporation, explained that their system, which offered optimum conditions for film drying, would extend the life of the film: "A few hours drying in heated air ages the film as much as several months in the tin cans will. . . . The hot air method robs the film of so much life."[37] With the temperature duplicating "natural conditions," this celluloid would not only last longer, Willat argued, but would always remain the same: "That is to say that the fortieth print from a negative will be exactly the same as the first."[38] Temperature control was also integral to the chemical processing of film, as the tanks of developing fluid were wired with a thermostatic device.[39] In the movie-appreciation text *Talking Pictures*, Barrett C. Kiesling gives thermostatic technology credit for the pictures that result: "Were that device to fail by two tenths of a degree . . . a love scene on the screen would lose its brilliance; carefully calculated lighting effects would lose their appeal to the visual sense."[40] Like the cold temperatures of the quantum computer, temperature control reorganized the media environment "to protect the delicate surface of the film on its journey through the laboratory," to stabilize the image itself, and to ensure that films appealed to the visual sense.[41]

The dependence of film production and manufacturing on temperature and the corresponding energy regimes required to manipulate the atmosphere helped prompt motion picture production to move from Fort Lee to Hollywood. In the winter of 1918, immense coal shortages brought about by World War I, along with the intense cold of that year, led to a series of regulations governing the sale of coal and the production of artificial heat. The film factories of Universal, Fox, Peerless, Goldwyn, Blache, Solax, and World (which together made up a significant percentage of the film installations in the United States at that time) felt the fuel shortage and the lack of heat "acutely": the industry news proclaimed that the coal famine had "affected the motion picture industry more keenly than any other branch of business."[42] Many industrial establishments were ordered to be closed. The lack of power, the cold, and the inability to battle these restrictions drove many studios west to capitalize on the atmosphere of Los Angeles. As the motion picture industry news commented in 1919, "None of the companies who have recently moved to the West counted on the closing of the war at such an early date, but rather believed that they would have the same trouble during the winter just closed."[43] Nonetheless, the failure to completely control atmospheric phenomena was one of the precipitating causes for the westward relocation of film production.

As air-conditioning was adopted across the media industries, it was not simply a technological advancement. The management of temperature materialized with a cultural imperative: to reduce the amount of error and noise in media representations, which would ensure that all moviegoers would view the same film. By expanding the thermal windows in which work could be done, these technologies accelerated the speed of film manufacturing. Take, for example, a film laboratory in Bayonne, New Jersey, around 1909, where climate control consisted of two coal stoves. In the winter, there was ice on the floor, and the men in the laboratory were able to heat the developing solution to 21.1°C/70°F, but they could develop only a few feet of film before the temperature dropped below 15.6°C/60° F. It took them a week to develop thirty-five thousand feet of film, working both day and night. By the 1930s, with atmospheric control a laboratory could develop that much film in just thirty minutes.[44] The expansion of cinema was underwritten by the expansion of atmospheric control in production facilities and by its capitalization on the thermal affordances of the California climate. At the same time, temperature control also made possible the synchronization of sound. Like the alignment of ink in print, the soundtrack negative had to be synchronized with the positive film. Engineers observed that without temperature control, variations in moisture content of the films would result in "changes of size and length . . . making synchronization exceedingly difficult, if not impossible."[45]

By 1931, air-conditioning was considered essential to film manufacturing. One set of engineers argued that while some theaters might choose not to cool their audiences and some soundstages might allow actors to "sweat through their scenes under the scorching heat of the incandescents . . . in the laboratory, if we fail to control temperature and humidity . . . we meet trouble and blocked production at every turn."[46] Yet even as these engineers discounted the significance of thermal control at the other sites of film production, temperature played a formative role in cinema's standardization across all of its sites. As was true for many media, the screening of films themselves produced heat, affecting the operators and projectionists. In 1907, nickel theater operators launched a protest against the San Antonio, Texas, electrical code, which required that the projection booth be lined with metal and have only a single opening to allow the projection itself. No ventilation was permitted. On busy days, operators were not be able to leave the booth, and with showings only ten or fifteen minutes apart, they had to spend time between showings preparing for the next film. Enclosed in the booth, it was nearly impossible for them to function. One reports: "The heat from the arc lamp raises the temperature to an unbearable degree and the metal-lined walls reflect the heat. It is like being in a red hot iron coffin without a breath of fresh air."[47] Temperature control was a means not only to standardize machinic and chemical processes but to optimize bodily processes of media operation.

In some cases, the management of heat conflicted with the management of other environmental processes, including sound and light. With the advent of sound film in the late 1920s, film studios were constructed to be soundproof, but the "methods of insulating against the flow of sound [operated] also against the flow of heat."[48] Describing a hypothetical example, one engineer wrote that in a seventy-foot-by-fifty-foot studio, the use of incandescent lights for a mere three minutes could raise the temperature by 72°F, from 65°F to 137°F. But air conditioners also generated noise, and vents used to dissipate heat circulated noise. This conflict also shaped the design of the radio and, later, television studios. A 1939 text on studio design described how the air-conditioning system, even as it "adds materially to the background noise in the studio," was able to combat the heat from the lights "not only for the comfort of the artists and producers, but to keep up the good condition of the musical instruments, [which] need repairing and tuning less frequently in studios that are air conditioned."[49] Sometimes the air-conditioning had to be turned off when the radio show was airing, otherwise the sound of cooling would overload the airwaves. This caused an increase in heat during the broadcast, especially for studios in urban locations, where broadcasts often happened in a room in the center of the building in order to buffer exterior noise. Studio production innovated a range of thermal technologies to carefully balance heat and sound, including soundproofing and adding canvas filters to air ducts.

Even with these conflicts, air-conditioning was as essential for studio production as for manufacturing and constituted a significant expense in the construction of new facilities. In January 1927, for example, the National Broadcasting Company (NBC) appropriated $400,000 for a new studio, outfitted with "an air-conditioning system said to be the largest except in South African mines."[50] The air-conditioning system was so notable it was featured in the studio's tours, with "the operation of mixers, coolers and filters described in detail."[51] Tour-goers had the opportunity to look through portholes into an eighty-five-thousand-gallon tank of cold water used for the system. In 1944, an NBC advertisement featuring the studio's massive artificial cooling system claims, "Mark Twain to the contrary, somebody *does* do something about the weather—makes it, in fact, just as he wants it—is able to forecast it months in advance" (figure 6.3).[52] Aligning with the contemporary discourses of thermostatic regulation, the NBC advertisement boasts that the company engages in "manufacturing made-to-order weather and processing it for string and percussion instruments, singers' throats and a host of performers who need the best atmospheric conditions."[53] At the top of the advertisement, a man alters a temperature control panel that is part of a broader media control panel. Here, heat and cold are just one other set of qualities of these industries' primary medium: air.

As the thermostat was being adopted in American homes, as viewers were encountering air-conditioning in cinemas, and as infrared photography was experiencing a boom, thermal conditioning was deployed across the mass media industries. This was driven by the cultural and capitalist imperative to achieve standardization across texts and viewing experiences and was defined by the attempt to sever media's thermal entanglements. As a result, the consistent production of media objects—in which each media text could operate without noise, error, or disruption—was fundamentally tied to a regime of thermopower, one that was grounded in the energy-intensive and often coal-powered maintenance of stable temperatures. The maintenance of even temperatures brought with it not only the expense of installing artificial cooling but the ongoing cost entailed in powering these systems.

the Weather
eats out of his hand

▶ Mark Twain to the contrary, some-body *does* do something about the weather—makes it, in fact, just as he wants it—is able to forecast it months in advance.

His is the job of keeping 4,200,000 cubic feet at NBC temperature ideal, dew point just what it should be, humidity never varying.

Humidity, that's the important thing in manufacturing made-to-order weather and processing it for string and percussion instruments, singers' throats and a host of performers who need the best atmospheric conditions for the best performances.

That's why NBC operates one of the most completely integrated air-conditioning systems in the world . . . 64 separate mechanical lungs so delicately coordinated that hun-

dreds of people may enter one studio without raising or lowering the temperature in the one next door.

This elaborate air-conditioning system helps keep actors and audiences comfortable and responsive, of course, but all that is secondary to the role it plays in contributing to better broadcasting . . . building a smoothness of presentation; a fidelity of instruments and vocal cords which contributes so much to the perfection of NBC's musical presentations.

* * *

Custom-built weather is just one of the many examples of the manner in which NBC strives for perfection in the little things of radio, knowing that the grand total of little things well done helps NBC maintain its leadership, helps make NBC "*The Network Most People Listen to Most.*"

They all tune to the

National Broadcasting Company
It's a National Habit

America's No. 1 Network

A Service of Radio
Corporation of America

FIGURE 6.3. An NBC advertisement foregrounds its "custom-built weather." National Broadcasting Company, "The Weather Eats out of His Hand," *Radio Daily*, August 1944, 63.

THE THERMAL MAINTENANCE OF MODERN MEDIA

Thermocultures—the cultural practices of temperature—not only prop up the manufacturing, production, and operation of media but also are fundamental to the ongoing maintenance of objects and infrastructures as media. Take, for example, early sound recording, which, like printing and film, depended on stable temperatures prior to and during production. Describing the Graphophone in 1908, Edward N. Burns remarked that the condition of its wax records was essential to expanding the export market for talking machines: "The wax blanks, the tools and every accessory, must be, and is, kept at a uniform temperature, so that when an original record goes into the 'bath' producing the copper shell, from which the 'master' is made, it emerges with the surface entirely free from oxidation, is smooth as glass."[54] Since the smallest imperfection could ruin a recording, Burns advised, "care must be exercised to protect the wax blanks from sudden or violent changes of temperature."[55] This concern was not limited to the export market—even in London, weather could "injure" records. The previous year, *Talking Machine World* reported that the cold weather had increased "the number of throw-outs at most of the factories."[56] While hot weather was not "feared as much now that improved cooling appliances have been installed in most factories," very low temperatures remained "disastrous."[57]

Like all media objects, wax cylinders remained entangled with heat and cold even after the initial moment of manufacturing: the cylinders expanded and contracted with changes in temperature. One Indiana phonograph cylinder dealer observed that this was exactly what made it difficult for phonograph players to accommodate different cylinder sizes. There needed to be some extra room on the mandrel, "for when the record contracts it will not go on to the mandrel to its original limit, and if expanded, vice versa. . . . Therefore if the record was made the full length of the mandrel and feed screw, one would be unable to play one or the other, depending on the temperature."[58] A variety of phonograph patents proposed inventions to address the problem of thermal entanglement. One from 1914 described a "movable record support" that could automatically loosen a record from its support, which was needed given that records changed in diameter depending on temperature conditions and could become tightly locked in place.[59] Another described a phonograph record holder that permitted "the record to freely contract or expand through temperature changes without danger of breakage."[60] Cylinder manufacturers proclaimed their products' resistance to expansion and contraction. US Everlasting Records boasted that its cylinders would neither expand nor contract, would fit any player, and were immune to "variant temperatures."[61]

The thermal entanglement of the phonograph player apparatus also generated problems. One difficulty existed with the talking machine's main spring, which as a tempered piece of steel was highly susceptible to temperature changes. One author cautioned, "Watch for chilled main springs," and warned dealers about delivering a machine on a cold day, since the springs might break when the new owner tried to play a talking machine immediately after installation.[62] One dealer went so far as to attach a special tag to the phonograph to help ensure that owners would allow their instruments to warm up for twenty-four hours. An advertisement for Vulcan Mainsprings alerted its customers that even a mainspring of the "best possible quality" was "sensitive to sudden changes of temperature"—even from a change as small as bringing one's gramophone into another room.[63] Other companies advertised that their components were unaffected by temperature changes.[64] In 1916, one company reported that it was specifically set to cater to the export trade market, with "machines so constructed as to withstand extreme temperatures."[65]

As phonograph cylinder size fluctuated with environmental conditions, expanding and contracting with the weather, those in the industry lamented that cylinders were "of no permanent value [because] sooner or later they always crack, owing to changes in temperature," and that "the proportion of discards due to cracking has been objectionably large."[66] Even disc records had a tendency to warp in part because of "the action of temperature," the head of one manufacturing laboratory complained.[67] Although technologies of temperature regulation were enlisted in standardization during production, the thermosensitivities of media persisted. Different materials—whether paper made from wood pulp, film with a nitrocellulose base, or the wax of cylinders—expanded, contracted, and reacted with adjacent materials depending on the climate. Quick or repeated movement between thermal states destabilized the media object. As a result, when phonograph cylinders, records, or magazines were left to bear the thermal fluctuations of natural environments, they lost many of their mediating capacities. Nitrate film combusted under even limited temperature shifts, and early film history was punctuated by explosions of overheated and agitated films. It is because of this fluctuation that relatively few phonograph cylinders and nitrate films exist today.

The preservation of media is, like its manufacture, linked to a regime of thermopower, where stabilization of temperature is directly connected to archival capacity. In a report for the Canadian Conservation Institute, the author writes that there can be no discussion of avoiding temperature altogether in media preservation, only a consideration of preventing "incorrect temperature."[68] The report includes a table that correlates thermal conditions to the approximate lifetime of a media object. Black-and-white photographic negatives on glass, produced in the nineteenth century, will remain usable for approximately seventy-five years in a hot room of 30°C/86°F but will last fifteen hundred years at 10°C/50°F. Newsprint and celluloid film will last only six months if left out in the sun, but at "normal" room temperature they will last a human lifetime. Magnetic media will last only fifteen years in a warm room of 25°C/77°F, but in cold storage at 0°C/32°F they will remain playable for six hundred years. Incorrect temperature, the report's author observes, is an agent of deterioration.

Because they depend on cooling technologies to maintain and preserve media, libraries and archives are key places where temperature's agential effects on cultural production are studied, albeit as an inevitable deteriorating force that disrupts the original form and function of media. Although there is certainly a threat of too-high temperatures that speed up chemical reactions and deform the media object, as well as too-low temperatures that fracture media's materials, most thermal practices of preservation are devoted to maintaining an even temperature, preventing natural fluctuations over the course of a day or a year. Architectures such as museums, through expansive technologies of climate control, become what Fernando Domínguez Rubio calls "objectification machines" that endeavor "to transform and stabilize artworks as meaningful 'objects' that can be exhibited, classified, and circulated."[69]

As a form of thermopower, the manipulation of media's production and storage environments is guided by cultural investment. In extending the lifetime of the media object, thermocultural practices attempt to arrest media in a particular form, whether one that mirrors the output of production or one that reflects a historical moment designated to be worth preserving. The object, standardized across environments and stabilized over time via thermal techniques, is endowed with an essentialized identity, yet this identity remains dependent on continued environmental control. The maintenance of objectness is underwritten by the cooling capacities of air-conditioning. These institutions and their thermocultural practices are the objectification machines of media.

This is of course only one example of the broader cryopolitical transformation of modern life, a transformation that harnesses artificial cooling to maintain the identity and objectness of all kinds of commodities. Fruits, vegetables, and meat are circulated around the world through a massive cold chain that prevents their deterioration and decay. Specimens and samples that compose the data of modern science are kept in cold storage. Problems in temperature control lead to rapid deterioration, blockage, and failure. One scientist recollects a moment in the 1960s in which a thermostat became "stuck," wiping out ten years of work in a single hour.[70] Ice core samples at the University of Alberta melted in the summer of 2017 due to a malfunction of the cooling system, destroying evidence important to climate scientists.[71] The materials of modern consumption are often chosen for their thermal properties and resistances. While authors have described the expansion of cryopolitical forms for contemporary food, science, and capitalism, few have paid attention to the massive cooling infrastructure that enables the manufacturing and maintenance of media, producing "a zone of existence where beings are made to live and are *not allowed to die*."[72]

THE COLDWARD COURSE OF COMPUTING

Media have taken a coldward course: over the twentieth century, artificial cooling shifted from a novel technique to a mandatory feature of media manufacturing, distribution, and preservation. Today's pressrooms employ intricate environmental regulations for optimal production: in the *New York Times*'s printing plant in Queens, mist sprays out in the air-conditioned reel room, where conditions are maintained at an ideal 22°C/72°F and 50 percent humidity.[73] Companies around the world offer custom air-conditioning systems that regulate the environmental conditions in paper mills (typically an environment "hostile to most equipment") as well as other media manufacturers.[74] As these cooling systems are installed across the media landscape, the form and content of media now depend on the operation of thermal systems: when the air conditioner shuts down, so does the operation. Just as standard temperatures were established across media manufacturing environments, they are also integral to the preservation of media. The American Society for Testing and Materials and the International Organization for Standardization, as well as many organizations dedicated to media preservation, set baselines for media's thermal environments. And just as the wax cylinder was eventually replaced with a much less fickle material, decisions about media's composition over the last century often reflect its thermal sensitivities and capacities. The quantum computer and the selection of niobium are only the latest iteration of this process.

Collectively, these examples illustrate the deep connection between media's materialization and a regime of thermopower that prioritizes both cooling and stability. This is nowhere more apparent than in digital media. Digital technologies are hypersensitive to thermal stimuli, more entangled with temperature than any media before them, and require cool temperatures and thermal evenness even more than newspapers, television sets, or videotapes. While for analog media, a shift to warmer temperatures tends to speed up decay (from a process unfolding over decades to one unfolding over years), for digital media, the lack of cooling can lead to instant disintegration. If analog media's thermal sensitivities tend to be analog, for digital media it tends to be binary: if it gets too warm for a second, the entire apparatus becomes nonfunctional.

As more and more elements are integrated in complex ways into media, each with its own reactivities and temperature dependencies, on the whole these technologies have become much more thermosensitive, requiring ever more controlled environments. In the case of digital media, temperature dependency has reached a point where air-conditioning systems initially developed for the standardization of materials are required for its very operation. As electrons are funneled through the varied components of digital systems, they give off heat that, if not managed, inhibits the machine's functionality. The extraordinary speed of conduction through pure materials generates an "incorrect temperature." Heat is managed to some degree by the incorporation of devices that redirect excess heat, such as fans and heat sinks. In large data centers, enormous cooling mechanisms are required to maintain the optimal temperature and ensure the stability of the computer's operation. The very objectness of any computational device is thus, from the moment it begins to compute, critically tied into strategies to maintain an even temperature.

As digital infrastructures have become more visible, there has been an increasing recognition—a thermoception—of the digital archive's dependence on cold and its vulnerability to warmth. The development of Facebook's Arctic data center and the Scandinavian server rush received extensive news coverage.[75] Thomas Pynchon's *Bleeding Edge* (2013) narrativizes the digital migration north in the United States and features a server located in the Adirondack Mountains. Timo Arnall's multiscreen installation *Internet Machine* (2014) features an air conditioner steadily humming in the background as the camera reveals the intricacies of internet infrastructure. Perhaps inspired by these cold reflections, in 2015 the television show *Mr. Robot* imagines that the data center's thermal dependency, rather than system code, is its most critical weakness. In the first season, the show's protagonist targets one archive whose loss would upend contemporary culture and capitalism: the debt records held by a massive conglomerate. Without these documents (which are stored digitally only), it would be impossible to verify obligation, and as a result, those living in a debt-ridden society would be freed from the whims of financial institutions. At the season's climax, the debt records are corrupted not by hacking into the record system but by manipulating data center thermostats. Capitalism, *Mr. Robot* imagines, is undone by its cooling system. The following year, another television show, *Westworld*, combined this narrative with the well-worn trope of *The Thing*—a monster set free by the melting ice. Here, artificially intelligent beings are released from cold storage following manipulation of the cooling system. In these popular imaginations, cooling is envisioned as a technology of control and interference with it as a means of releasing the repressed. Such narratives make sense precisely because their audience is already primed by long-standing colonial tropes that situate the too-warm tropics as undoing the foundations of Western civilization. Progress, in both news and imaginative representations, is ensured by coolness. These discourses help to naturalize a popular sense of the coldward course of computing.

Such fantasies of thermal disruption have yet to be realized, even if computer engineers have begun to identify temperature and thermoregulatory systems as potential sites of cyberattack. Cybersecurity experts, describing something along the lines of *Mr. Robot*'s plotline, suggest that targeting temperature sensors would be an ideal mode of attack for hackers who don't want to get caught.[76] But intentional attacks on HVAC systems are rare. Thermal problems are triggered far more often by human or machine error. A simple human mistake, changing the register from Fahrenheit to Celsius, heated one server room to over 37°C/100°F, melting drives in the process.[77] Even when it is not a specific target, the cooling system is a vulnerable infrastructure that can compromise archival practices. The oscillating temperatures of global warming only increase this vulnerability. In one notable example, the melting of the permafrost around the Svalbard Global Seed Vault flooded the entrance, although it didn't affect the seeds inside.[78] Like all archives, the vault was constructed with an assumed thermal baseline. When temperatures exceed predicted levels, cooling systems do not always function as intended. In addition, temperature regulation itself is often one of the primary energy expenses of media's archives. Fossil fuels are burned to produce a thermally consistent environment that stabilizes archival material, which is then made more vulnerable by the variable temperatures of a changing climate, necessitating more investment in the technologies of cooling.

In light of this, research in the data center industry has focused on data centers' energy consumption, the potential to reuse their "waste heat," and the potential "free cooling" offered by the air and water around them.[79] This research has stimulated architectural and infrastructural innovations that redirect excess heat into businesses and even homes of urban dwellers—producing what Julia Velkova describes as a new "computational traffic commodity."[80] In Seattle, the heat generated by the Westin Building Exchange, an internet exchange with data center capacity, is "recovered" by

Amazon, ported through a "heat exchanger," and funneled to the company's Denny Triangle campus.[81] The heat-recovery system not only warms the bodies of Amazon employees but also provides the company with an environmentalist narrative, complete with pictures disseminated via social media, about its "contribution to Seattle's goal of carbon neutrality."[82] In Haifa, Intel uses the waste heat generated by its IDC9 data center to heat the building's offices and warm its water. Data centers heat a community pool in Switzerland, the newsroom of the *Winnipeg Sun*, and a greenhouse and botanical garden in South Bend, Indiana, among many other sites.[83] Given the extra infrastructure this requires, setting up heat recovery is not always feasible for smaller data centers and thus tends to be a practice of companies with significant power in the digital infrastructure industry. Moreover, the transfer of heat is difficult—this is a distribution problem as well as an economic one, and regulation poses barriers as well, since in some places it is illegal to sell excess electricity to one's neighbors.

While data centers' waste-heat recovery is a complex infrastructural solution that reroutes and capitalizes on thermal expenditure, a more common strategy for sustainable system development has been simply to warm up the data center. Increasing the temperature only a few degrees is regularly touted as a choice to save money, energy, and the environment. The American Society of Heating, Refrigerating, and Air-Conditioning Engineers (ASHRAE) publishes thermal guidelines for data-processing environments, and over the years it has raised the highest temperature that the servers can withstand. Researchers at the University of Toronto, using Google, Los Alamos National Laboratory, and the SciNet Consortium as their subjects, argue that data centers could turn up the heat and contest the assumption that failure rates of electronics double with every 10°C/18°F increase. Variability in temperature, they state, is a more critical concern than the amount of heat.[84] Large-scale operators have become trailblazers in changing the norms: Google, for example, runs some facilities at 26.7°C/80°F.[85] In the digital infrastructure industry, there is a growing acceptance of increasing the temperature. But the choice to cool and the degree of cooling are culturally inflected decisions that concern security and safety. As one data center analyst argues, the reality is that decisions about cooling are often the result of "exaggerated safety margins" that have resulted in "a historic practice of overcooling facilities."[86] Not all data center operators heed the call to warm up, hesitant to compromise their equipment with too-high temperatures.

The rerouting of computer-generated heat, the dependence on extensive cooling systems, and the management of data's environments are the latest iterations of the coldward course of media. They are also evidence of the continuing connection between digital systems and thermostatic operation. The thermostat has long been influential in the conceptualization and development of control systems, early cybernetics, and automation. Indeed, Norbert Weiner's *Cybernetics* describes the "ordinary" thermostat as a foundational example of automatic controls, "a purely mechanical feedback system" in which "no human element intervenes."[87] As cybernetics traveled to other domains and disciplines, the thermostat was used to help to explain its theories.[88] In turn, early thermostat control systems were described as data centers. In 1956, Honeywell reported that its complex electronic temperature control panels, "call[ed] Data Centers," were regarded "as having great potential."[89] Building on their experience in thermostat design, Honeywell would go on to be foundational in early computer systems development.

Today, as thermal sensing devices translate the world and its heat into thermal data, this, too, is stored in data centers. All of the digital temperature control devices described in earlier chapters—from smart thermostats to agricultural thermal imagers—generate massive amounts of information to be processed in algorithmic operations. The attempt to record and predict future fluctuations on the basis of past data is integral to the adaptive subject, networked temperature, and thermal microcasting. As a result, seeing by heat and modulating everyday environments using digital systems, even for environmental purposes, scales up the very emission of heat into the world, producing unreadable and as-yet-unpatterned fluctuations of climate, repositioning more bodies, and necessitating more thermal media to register and account for these heat shifts. In other words, the more networked thermal media is used to sense and reshape heat, the more heat there is to see, the more climate changes there are, and the more important such technologies seem to become. As transmission accelerates and data accumulates, generating more and more heat, the need for a cooling apparatus to stabilize it only grows. This is media's thermal loop.

Computation today, digital and quantum, is embedded in a regime of thermopower, where the entanglement of media and temperature is shaped by social practices and cultural values. These can be easily seen in the many unquestioned assumptions in the design of heat and thermal control. Within the infrastructure industries, the process of grappling with climate change and energy costs has not involved diminishing or reducing the amount of data or transmission—whether high-tech or low-tech, strategies almost always focus on altering computation's thermal contexts rather than scaling back production or distribution. Outside of the data center industry, alternative approaches have emerged, with some cultural heritage institutions working to reduce the amount of energy they use and some archivists acknowledging the inevitability of loss and the costs of commercial data management—especially with commercial platforms' insistence on redundancy and multiplicity. Moreover, the choice to use energy-intensive systems to offset heat, rather than architectural modifications or spatial reorganization (say, in an imagination of distributed data centers), reflects the industry values of centralized systems, privacy, security, and safety.

The coldward course of computing and the increased thermal entanglements of computational media mean that the environment—and thermal capitalism—is playing a greater role than ever in the history of media and communication. While there has been significant news coverage of the northward migration of servers and data centers, much less attention has been given to the problems of developing computing infrastructure in tropical climates. Historically, internet infrastructure has been concentrated in the United States and Europe, and many internet exchanges, data centers, and cable landing stations remain hubs there even as the network expands around the world. However, as one facility services firm observes, the need for additional data centers in remote locations has become "more acute," and tropical locales continue to "offer a host of unique challenges."[90] They have to be "more aggressive with cooling," and since they require increased power usage they have lower levels of efficiency: climate makes them "efficiency-adverse locations."[91]

The cost of cooling and maintenance is higher for data centers in warm areas, especially tropical areas with significant moisture. Singapore, for example, is a hub of internet infrastructure in part because of its geographic location, colonial history, and position as a hub in the global undersea cable network. It possesses many of the ideals for network infrastructure—connectivity, a relatively stable electrical grid, the lowest corporate taxes in the Asia Pacific region—and for that reason is the location of at least eighty data centers, earning it the title of "Asia Pacific's Data Center Capital."[92] The data centers in Singapore include not only local start-ups but global chains such as Digital Realty, Telehouse, Iron Mountain, and Equinix, among others. To be globally competitive, however, costs more energy and money in Singapore than in many of its northern counterparts. In the industry, it is known as "a punishing place to build," and designers cannot rely on techniques of evaporative cooling, for example, due to the humidity.[93] Reflecting on their country's network infrastructure, Singaporean scholars argue that the high heat and humidity are "a major disadvantage" and observe that the cost of running data centers amounted to 6.9 percent of the country's electricity usage in 2012.[94] As a result, Singapore has been the site of several innovative proposals in data center design, including a proposal for underground data centers as well as "the world's first tropical data center," designed to function at much higher temperatures.[95]

The problem of higher cooling and energy needs not only puts tropical data centers at a practical economic disadvantage but marginalizes such sites in the marketing of data centers and in industrial perception of network viability. One particular difficulty in this regard revolves around a metric for assessing data centers. Data centers are evaluated for energy efficiency using a metric referred to as PUE (power usage efficiency). This metric was initially designed by the Green Grid—an industry consortium of large global technology companies primarily based in the United States, the European Union, and Japan—to improve data centers' environmental and economic efficiency. It was later published as a global standard, ISO/IEC 30134–2:2016, and briefly included in the ASHRAE standard for data center efficiency (ASHRAE 90.4) before being replaced by a more granular metric. Essentially, PUE refers to the ratio of the total energy used by a data center to the amount of energy delivered to the actual information technology equipment (not including any cooling or power for the facility itself). This metric is not only used by data center managers to market their facilities as environmental or efficient but by local regulators, especially in the United States. In 2016, the city of Beijing issued a ban on data centers with a PUE of 1.5 or above.

One of the problems with PUE in tropical areas is that this metric does not evaluate data centers based on how much energy they consume but rather ranks them based on how little energy they use to buffer their equipment from the environment—to mitigate thermal entanglement. In this case, hotter areas, which cannot use ambient cooling, will inevitably have a worse ratio than those with "free" cooling. One analyst comments, "It did not take long for the cooling system to be considered the greatest enemy of PUE."[96] In the case of Singapore, for example, the "best-in-class" multitenant data centers have a PUE of 1.44 at the lowest, compared to a rating of 1.185 for a similar data center in Nevada.[97] Industry materials, in turn, emphasize the "free cooling" of Europe, North America, and Japan. The Green Grid published "free cooling maps" in 2009 that depict the ambient temperature as a resource to be leveraged alongside tax benefits, electricity costs, and other site-specific features.

Climate does not simply determine data center location (and, by extension, the location of the internet). As those in the industry reiterate, climate

TABLE 6.1. Top Markets for Data Centers and Mean Annual Temperature

Rank	Location	Mean Temp °F	Rank	Location	Mean Temp °F
1	London	50.5	26	Brisbane	70.5
2	Frankfurt	51.1	27	Vienna	50.7
3	Amsterdam	50.4	28	Madrid	59
4	Washington, DC	58.2	29	Minneapolis	46.1
5	Hong Kong	73.9	30	Boston	51.7
6	Los Angeles	65.4	31	Montreal	44.2
7	Bay Area (CA)	58.2	32	Munich	46.4
8	Dallas	66.2	33	Manchester	50.9
9	Sydney	63.9	34	Brussels	50.9
10	Paris	54.1	35	Milan	55.4
11	Chicago	49.7	36	Dublin	49.6
12	Singapore	81	37	Warsaw	47.3
13	New York	55.3	38	Perth	65.7
14	Atlanta	62.6	39	Dusseldorf	50.4
15	Toronto	48.9	40	Berlin	50.5
16	Jersey City	52.7	41	São Paulo	66.6
17	Tokyo	59.7	42	Mumbai	80.8
18	Denver	50.7	43	Houston	69.4
19	Seattle	52.4	44	Kansas City	54.5
20	Miami	76.9	45	Copenhagen	48.4
21	Melbourne	59.2	46	Rotterdam	49.3
22	Phoenix	75	47	Philadelphia	55.8
23	Moscow	42.4	48	Auckland	59.4
24	Stockholm	43.9	49	Sofia	51.1
25	Zurich	48.7	50	Las Vegas	68.5

The ranking of data center markets is from Cloudscene, "London, Frankfurt and Amsterdam Revealed as the Top Three Markets to Colocate in 2018," January 31, 2018, https://cloudscene.com/news/2018/01/markets-to-colocate. The data on mean temperature is from Wikipedia, "List of Cities by Average Temperature," accessed April 4, 2021, https://en.wikipedia.org/wiki/List_of_cities_by_average_temperature.

may be an influence, but the concentration of internet infrastructure in particular places is also a result of local economic contexts, availability of power (since cooling accounts for 30 to 40 percent of data center energy consumption), connectivity, and, "very importantly, politics."[98] The internet maps onto a grid of resource potential that is as shaped by colonial and corporate investment as it is by climate. As a result and in part because of its connectivity, Singapore has more data centers than Denmark and Finland combined, even with these countries' cool climates and access to renewable energy. But Singapore also attracts investment precisely because many of its immediate neighbors face historical, political, and thermal challenges to internet infrastructure development. Moreover, the emphasis on "free cooling" and the "insane" and "inefficient" practice of locating data centers in hot climates shape the perception of the viability of these locations and thus their capacity for development.[99] Warm facilities are perceived as less secure and less reliable. The selection of sites is often shaped by internal industry discourse. Take, for example, Cloudscene's recommended list of hubs for network exchange and development (table 6.1). Singapore ranks twelfth on the list, and only one other location has a mean temperature above 26.7°C/80°F. The other locations are largely located in cooler climates in Europe and the United States.

On the one hand, the coldward course of computing, even with innovations such as the use of solar power for data centers in tropical locations, means that the internet—as it is currently designed—requires that locations that aim to produce, transport, or maintain media must develop the capacity to maintain stable temperatures. The push toward ecological computing, especially in efficiency metrics, continues to privilege not only the Western domination of internet infrastructure but also the hyperscale operators such as Google and Facebook that can afford to manipulate temperature at such an expansive degree and using artificial intelligence. The privileging of security, redundancy, and reliability and the tying of these analytics to the cold help to striate internet infrastructure and to ground it in locations where temperature and politics appear to remain "friendly." As an alternative, there are many other ways of approaching the problem of overheating that don't leverage climate: decreasing the amount of communication, spatially distributing computation, and investing in alternative energy, to name a few. Such options remain marginal, if considered at all, as our contemporary mediascape remains embedded in a regime of thermopower that prioritizes precise thermal management which, in turn, is justified by the threat of climate change.

THE EXTENSIONS OF MEDIA

An extension is a part that has been added on to something, increasing its size or duration. An extension prolongs, expands, and amplifies. Marshall McLuhan defines media as that which extends the senses, increasing one's perceptual reach and communicative capacities.[100] History will be recorded because books outlast the human life span. Television enables sight across vast distances and into realms uninhabitable. Technologies such as thermometers amplify the ability to sense heat and cold. Networked thermostats, paired with heating and cooling systems, enable the control of heat and cool from afar. Indeed, Paul Wishart, president of Honeywell in 1956, before McLuhan's media theoretical writings, argued, "Everything man does, in terms of technical progress or mechanization, is an effort to extend himself beyond his own limitations."[101] While the telephone and radio, Wishart explained, "simply projected speech and hearing," the thermostat is "a super-extension of [a person's] nervous system—being faster, more accurate, applicable to almost all his functions and completely automatic."[102] Like other media, thermostats and air conditioners extend the capacity to communicate.

This chapter shows how, since the beginning of the twentieth century, thermal technologies have been a primary means by which other forms of communication are prolonged, expanded, and amplified. Thermal technologies extend the scale of production of traditional mass media, its longevity and duration, and its expansion around the world, from the export market of early phonograph cylinders to the contemporary development of internet infrastructure. As this chapter reveals, the development of modern thermal technologies has been as critical to the development of industrial products, including almost all traditional media technologies, as to the comfort of human users. Thermal technologies are not only extensions of people; they are the extensions of media.

Thermal technologies, especially cooling technologies, extend media in particular ways. They standardize media so that it can be made uniform and its production can be scaled up. They hold media in suspension, maintaining the objectness of media objects and attempting to defer their inevitable death and degradation. And they enable the spread of media around the world, even as they tether media to existing regimes of thermopower shaped by a colonial geography in which the "warm lands" face increased challenges to maintaining their place in the mediascape. Dwelling on these techniques reveals the tensions between media access, equity, and environmentalism that are built into the history of modern media. While the narratives about computing's coldward course draw upon a long history of colonial knowledge to naturalize this move north, it is critical to understand that the design of media infrastructures and systems is a deliberate social production. Media have been built in a way such that their extension into warm climates and tropical locations will cost more—economically and ecologically. They could be designed otherwise: to degrade over time or to withstand thermal variation.

Many of these thermal technologies have been sold as ecological techniques that reduce the amount of waste generated in media production and circulation. Early adopters of air-conditioning in various sites of media manufacture lauded the efficiency of the temperature-controlled factory because fewer materials had to be discarded as a result of "defects." The stabilization of media in archives and other thermally controlled environments keeps them from degrading and thus becoming waste. What these offer, however, is a shift to energy-intensive cooling mechanisms that substitute the waste produced by fossil fuels for the waste of media objects.

An attention to media's cryopolitics and to these invisible thermal regimes also exposes the need for cold as *the* critical vulnerability of media, the one that could corrupt all of our data in an instant. The imagination offered by *Mr. Robot*, in which capitalism is undone by a thermostat, is not far from reality. An alteration of the thermal contexts of media and communications, an alteration of their invisible infrastructures, would corrupt media history, would capsize digital communications, and would disrupt the production of thermally dependent media objects at scale. This is the threat of climate change to contemporary mediation. Far more than rising sea levels or storms, volatile and unexpected temperatures are likely to exceed the safety limits of infrastructures that facilitate media exchange and store media objects. But this will not be a great equalizer of media infrastructure or access. The first media that are likely to be disrupted are those without backup systems and redundancy in their thermal management. Media at the thermal margins, in the places where it is already more difficult to sustain signal traffic—economically, politically, and environmentally—will be even more precarious.

In August 2016, temperatures soared in the Yamal Peninsula, at the far western end of the Siberian Arctic. The heat reached a record high of over 32°C/90°F. Foreign reporters compared the area to the tropics. A slice of the permafrost melted, thawing reindeer bodies that had long been frozen. Bacteria once immobilized by the cold were released. Anthrax quickly spread among the area's reindeer population, one of the largest in the world and the primary life-support system for Yamal's nomadic inhabitants. Infection spread to human bodies and a boy died.

This chain of events garnered the attention of international journalists. The reanimation of bacteria reanimated coldsploitation stories about tropical threats at the poles. This gave traction to scientists' warnings that ancient viruses might surface as the ice retreats. In these narratives, melting is a crisis event, not because of the floods it will precipitate but because it liberates something previously trapped in the ice. Temperature change is a catalyst that speeds up circulation and ultimately accelerates structural collapse.

As the permafrost melted in Yamal, the softened tundra, the unimpeded wind, the corpses of reindeer, and the bodies of people all became viral media. Ecologies were activated as a set of transmissive surfaces: a hot zone. The Russian government attempted to halt these transmissions by relocating people, killing reindeer, and spraying massive amounts of disinfectant that would prohibit growth for years to come. The extermination of the animals also functioned to devastate the nomadic ways of life harnessed to them. The perpetrators of thermal violence once again deferred accountability to the environment.

In the months that followed, a second process of melting gained far less attention. With the warmer temperatures came a massive amount of rain. Climate change is characterized by rapid thermal oscillations more than steady warming, and the area was set to plunge into the frigid cold. Melting does not simply release deviant elements. It often homogenizes the substance that remains. In the Yamal Peninsula, melting not only released bacteria in reindeer carcasses but also freed air trapped in between molecules of water. Water molecules became more tightly packed, a process called densification. When the cold hit, the ice became a hard barrier, more difficult for reindeer to break through in order to eat. This process, described by researchers as a "rain-on-snow event," had happened before. In 2013, nearly one-quarter of the peninsula's 275,000 reindeer starved. And for the people who depended on these reindeer, that loss was far more catastrophic than the release of the bacteria by itself.

Melting is a phase transition, a process by which a substance shifts between solid and liquid states, typically brought about by changes in temperature and pressure. As Janet Walker suggests, climate change itself is "matter out of phase."[1] While phase transitions include evaporation and boiling (liquid to gas), condensation (gas to liquid), and freezing (liquid to solid), climate change discourse largely focuses on the transition from solid to liquid. Melting ice, Cymene Howe and Dominic Boyer show, has been a key "climatological signal" that has rendered global warming visible.[2] Scientists have measured polar melt to index thermal transformations. Films such as *Chasing Ice* (2012) have visually documented the melting of glaciers in order to ground statistical charts and diagrams. News stories have tracked incidents such as the release of "zombie" viruses. The companions to the images of melting are images of coastal flooding, which collectively generate an imagination of water displaced from the poles to islands. This arc of climate change (heating causes melting causes flooding) provides a structure to popular books like *A World without Ice* (2009) and *After the Ice* (2009).[3] What is solid will become liquid, and eventually life will be submerged under water. It is a unidirectional process that mirrors the movement of a warming world. And in almost all of these representations, the melt is the crux of a crisis narrative.

The perception of melt as a crisis, as a marker of a singular and critical transition, is inflected by Western theory that regularly uses phase transitions to describe social transformation. In the nineteenth century, theorists of cultural and ideology adapted the language of thermal machines around them to describe the social changes they were witnessing. For Karl Marx and Friedrich Engels, the bourgeois epoch is distinguished by the constant revolutionizing of production, in which "all fixed, fast-frozen relations, with their train of ancient and venerable prejudices and opinions, are swept away, all new-formed ones become antiquated before they can ossify. All that is solid melts into air, all that is holy is profaned, and man is at last compelled to face with sober senses his real conditions of life, and his relations with his kind."[4] The thermal transition from solid to liquid to gas, and the corresponding genealogy of the "meltdown," is an apt metaphor for the dissolution of apparently solid structures. But phase transitions are also integral to capitalism itself, and their capacity to alter the homogeneity and form of substances makes thermal capitalism possible. From the nineteenth century to today, phase transitions have been a part of the production of almost all industrial technologies. And they are critical, as the previous chapter shows, to the creation of all media technologies.

But even in the composition of industrial materials, thermal transitions are rarely unidirectional.[5] Multiple meltings and solidifications are essential to the circulation of media on a mass scale, to their transformation from variegated material to homogeneous media. These processes tether media production, like many industrial and postindustrial processes, to a fossil fuel–based energy regime and thus undergird the ongoing warming of the earth. So too, as the previous chapter illustrated, does the arresting of thermal change through cooling. Thermal technology, as Mél Hogan and Sarah T. Roberts point out, "offers the solution to the melt, but invariably—when failed or failing—leads it to meltdown."[6]

In contrast to the crisis narrative of overheating and melting, *Media Hot and Cold* documents the complex cultural effects of nonlinear thermal processes that will become pervasive in the wake of climate change. Although from the perspective of geologic time, the melting of polar ice represents a loss of a particular form of environmental stability that is unlikely to return, climate change involves a multitude of meltings and recrystallizations of varied substances occurring at different rates across the globe. People, bodies, lives, and media will remain thermally entangled. In fact, as in the case of the "rain-on-snow event," phase transitions will often cause densification and barriers that impede alongside the production of viral media. The effects of melting and solidification vary across environments, terrains, topographies, and cultures. Climate change isn't simply a decomposition but a transformation of substrates that support movement and exchange. In other words, climate change alters the capacities of the world as a medium.

Over the past two decades, many academic fields—including environmental media studies, environmental communication, and the environmental humanities, as well as numerous others—have sought to make sense of climate change. Researchers have tracked how wildly oscillating temperatures are made legible in the news, how photographers have projected a vision of climate, how ecological disasters have been modeled in cinematic form, and how soundscapes have materialized in ecological art. Given the scale of climate change, much of this research has correspondingly focused on large-scale transitions, such as melting glaciers, and highly perceptible effects, from intense hurricanes to massive droughts. For some researchers, the immense scale of globally interconnected changes produces a sense that climate change is unknowable, unrepresentable, and beyond human knowledge.

Media Hot and Cold points readers in the opposite direction—toward the embodied, the visceral, the everyday, and the social. Climate change is not invisible. It is everywhere present, perceptible, and affective. Situated knowledges of temperature and vernacular forms of thermoception exist in many people's everyday lives, whether they work under a radiant sun, in a sweatshop, or in an overchilled distribution center. And these thermal shifts are perceptible in the new food systems that have emerged and the thermal storage systems that support them. They are perceptible in the notes about heat and cold in prison writings. They are perceptible in thermostatic interfaces. They are perceptible in images of infrared cameras.

Media Hot and Cold attunes readers to these moments and sensibilities, which are held everywhere in people's bodies. These are conditioned not only by the development of digital thermocultures and by the thermal media of the early twentieth century but by a much longer history of thermal manipulation that is bound up with colonial, racial, and gender dynamics. As Kyle Powys Whyte has shown, counter to Western narratives of climate change as a dystopian and apocalyptic event, colonization itself has long harnessed climate change as a technique of thermal violence against Indigenous peoples.[7] Although the Anthropocene names a "break in the understanding of time and geologic epoch," Kathryn Yusoff writes that "it is also a material cut into bodies: real, actual, specific, vulnerable bodies; bodies of those that do not get to count as fully human in the current biopolitical order; bodies of earth, bodies of nonhuman organisms, social and geologic bodies that matter."[8] These cuts are often materially made through thermal media.

While people have long been conditioned to see images, hear sounds, and feel interfaces as media, sensing heat and cold as forms of communication runs up against a deeply ingrained sense of thermal objectivity. For this reason, this book turns toward metallurgy and generates a sensory grammar for thermoception. In this sense, it borrows not only a title but a tactic from Marshall McLuhan, whose words produced a tactile vision of the media landscape. In McLuhan's books, language and images do not objectively describe the rapidly shifting media environment—they evoke such transitions. Hot and cold media, in part due to McLuhan's writings, are already embedded in media studies' sensory landscape. Like the words of so many other theorists, the dominant assumptions in these sensory landscapes enfold latent power relations, naturalizing whiteness, colonialization, patriarchy, cis-gendered bodies, and heterosexuality.

Theory does not operate in an abstract conceptual landscape but in and on bodies. *Media Hot and Cold* attempts to work on the senses, to attune readers to the air, the spectrum, and bodies as means of communicating heat and cold. In this, the book is indebted to a lineage of theory on the elemental work of mediation. For decades, feminists have been drawing into relief how the body itself has been constructed as a medium.[9] And in their seminal book *Life after New Media*, Sarah Kember and Joanna Zylinska articulate mediation itself as a vital process. Most recently, John Durham Peters advances a philosophy of elemental media, illustrating the many ways that the environment itself functions as a form of mediation.[10] While McLuhan's language offers a tactile vision that naturalizes technical systems as a colonial environment, the evocative language and concepts of more recent works—and this book—inverts the process: the goal is attunement to the technicity and mediation of what we assume to be "natural."

Counter to the dominant sensory regime of thermal objectivity, *Media Hot and Cold* attempts to alter the reader's sense of temperature. To generate a sense that one's body is an emitter, a receiver, and a conduit for all kinds of thermal communications. To generate a sense that bodies are already structured by industrial systems that seek to profit and control from thermal sensation and emissions. To show the various media industries that reshape thermal forms, from the air-conditioning industry to the film industry. The topography of this landscape, its bodily transmitters and receivers, and the structural shifts brought about by changing climates can be felt as the work of thermopower. Proliferating concepts is a means of distributing critical thermoception as a sensory mode, one that can attune readers to the felt reverberations of a changing climate.

Alongside a sensory formation, *Media Hot and Cold* articulates a politics of thermal media. As the environment is increasingly mediated by thermal technologies and as large-scale ecological shifts offer opportunities for thermal violence in an ever-expanding set of locations, there is a need to establish new baselines and more complex models for thermal rights. These must take into account the dynamism and entanglement of people as thermal beings, as well as the ways that the body's thermal capacities can be shifted not only through a change in temperature but through metabolic, hydrologic, and other circumstances. Yet as I show in the chapter on thermal violence, by themselves, thermal rights are insufficient.

Mitigating climate change in equitable ways requires thermal autonomy: the expansion of people's capacity to shape the circumstances of their own vitality. There can be no thermal autonomy achieved through today's digital systems, as such systems inherently redistribute control and corporeality across corporate networks. There can be no thermal autonomy when there is incarceration, as carceral practices inherently limit peoples' capacity to govern their vital activity. Thermal autonomy is not a disavowal of connection—it stems from the recognition rather than the refusal of entanglements, thermal and otherwise. It starts with social reorganizations that offer more just means of engaging one's world and, as a result of this, one's temperature. Social, cultural, and media politics often carry with them a latent politics of temperature.

Critical to these reorganizations are thermal media: air conditioners, heaters, architectures, fireplaces, infrared technologies, ice, building insulation, and representations of heat and cold, among many others. Addressing changing climates is a media issue. Without access to, control over, and local governance of thermal media systems, there can be no accountability for thermal violence. Without the establishment of local and regional norms of thermal media languages and operation, there can be no systemic calibration to local thermocultures. Without autonomous thermal media production, there can be no capacity to locally mitigate the temperature effects of climate change. Digital systems are a significant threat to the environment not only in their energy costs but also in the way that they fundamentally preempt possibilities for people to establish their own baselines for temperature. The questions from here are, What thermal technologies, ethics, policies, and spaces allow for thermal autonomy? Are there forms of thermal localization that can short-circuit networked systems? How might communities harness thermal media in response to the many meltings, solidifications, crystallizations, and ecological shifts in the century to come?

Preface

1. Here I have been inspired by and extend the work of Jamie Skye Bianco, "#inhabitation" (multimedia/digital storytelling), *CCCC Online* 1, no. 1 (spring 2012), https://cccc.ncte.org/cccc/ccconline/v1-1/bianco.

2. Osman, *Modernism's Visible Hand*, 23.

Introduction

Portions of this chapter were previously published as "The Materiality of Media Heat," *International Journal of Communication* 8 (2014): 2504–8.

1. Klein and Nellis, *Thermodynamics*, 1.

2. See Eric Klinenberg's *Heat Wave* for a description of social infrastructure's redistribution of heat effects.

3. Lauren Berlant and Lee Edelman, *Sex, or the Unbearable*, 25.

4. As Elena Beregow suggests, thermal intersubjectivity is made possible "through the melting and mingling of bodies." Beregow, "Thermal Objects," 2.

5. One notable site where thermal communications (and thermodynamics) have been influential is in the development of cybernetics and associated media theories. See Dylan Mulvin and Jonathan Sterne's special issue, "Media Hot and Cold," in the *International Journal of Communication* in 2014; and Ernst, "Time, Temperature and Its Informational Turn."

6. Here I am inspired by the work of Dora Silva Santana on "tactile grammars" (Santana, "Trans* Stellar Knot-Works"), as well as by Melody Jue and Rafico Ruiz's articulation of the elements as heuristics (Jue and Ruiz, "Thinking with Saturation").

7. The gendering of temperature is both a social and a physiological phenomenon. Temperature is not only differentially felt according to social activity and clothing; it's also affected by hormonal and metabolic processes.

8. Hobart, "Cooling the Tropics."

9. Douglas, "Environments at Risk," 207.

10. Hess, "Standardizing Body Temperature."

11. Here I rely on both classic descriptions of communication as the processes of transmission and reception for information and on more recent expansions of media theory that extend communication beyond human-to-human messages to environmental phenomena that provide "conditions for existence" (see Peters, *The*

Marvelous Clouds, 14) and to the "interlocking technical and biological processes of mediation" (see Kember and Zylinska, *Life after New Media*, xiiv). I also invoke Lisa Gitelman's foundational definition of media as "socially realized structures of communication" (*Always Already New*, 7).

12. Fennell, *Last Project Standing*.

13. Day, "98.6."

14. Mbembe, "Necropolitics." For a discussion of thermopolitics, see Daggett, *Birth of Energy*; and Clark, "Infernal Machinery."

15. Here I join many other scholars, who have in recent years have extended Foucault's description of biopower to account for environmental processes. Energopower, Dominic Boyer writes, can help scholars to grapple with some of the "shockwaves set off by overuse of carbon and nuclear energy" and account for the way energy systems "have shaken the foundations of contemporary biopolitical regimes." Boyer, *Energopolitics*, 12. See also the formulation of ontopower in Massumi, *Ontopower*, and geontopower in Povinelli, *Geontologies*.

16. Royston, "Dragon-Breath and Snow-Melt"; Lara, "Affect, Heat and Tacos"; Vannini and Taggart, "Making Sense of Domestic Warmth"; Allen-Collinson and Owton, "Intense Embodiment"; Potter, "Sense of Motion, Senses of Self"; Classen, *Worlds of Sense*.

17. Ong, "Warming Up to Heat," 7. Historians of architecture, in particular, have developed numerous lines of inquiry into the relationship between temperature and culture. In *Modernism's Visible Hand*, Michael Osman tracks the ties between thermostatic systems, cold-storage warehouses, and regimes of regulation.

18. Barber, *Modern Architecture and Climate*, 9.

19. Ackermann, *Cool Comfort*; Adams, *Home Fires*; Chang, *Inventing Temperature*; Cooper, *Air-Conditioning America*; Middleton, *A History of the Thermometer*; Osman, *Modernism's Visible Hand*; Woods, *Herds Shot Round the World*.

20. Hutchison, "Journalism and the Perfect Heat Wave"; LeMenager, "Living with Fire"; Peters, *The Marvelous Clouds*; Ruiz, "Phase State Earth."

21. See, for example, Ruiz, Schönach, and Shields, "Introduction: Experiencing after Ice"; Schönach, "Natural Ice"; Radin and Kowal, *Cryopolitics*; and Radin, *Life on Ice*.

22. For a discussion of the thermal and the production of art objects, see Rubio, *Still Life*, 22. See also Elena Beregow's special issue of *Culture Machine*, "Thermal Objects," which assembles a range of inquiries into the artistic and performative dimensions of thermal objects.

23. Hulme, *Weathered*.

24. See Allen-Collinson et al., "Exploring Lived Heat"; and Vannini and Taggart, "Making Sense of Domestic Warmth." Alex Nading has suggested that ethnography can "break down the metaphorical and social forms of insulation that make heat seem sometimes global, sometimes invisible." Nading, "Heat."

25. Horn, "The Aesthetics of Heat"; Venkat, "Toward an Anthropology of Heat."

26. Fretwell, "Introduction: Common Senses and Critical Sensibilities."

27. McLuhan, *Understanding Media*.

28. Vannini and Taggart, "Making Sense of Domestic Warmth."

29. Serres, "The Origin of Language," 73.

30. Günther Selichar, *Screens, Cold*, 1997–2003, series, in *Offene Grenzen*, exhibition, Künstlerwerkstätten, Lothringerstraße, Munich, 1998.

31. Alter, Koepnick, and Langston, "Landscapes of Ice, Wind, and Snow."

32. Wollen, "Fire and Ice," 78.

33. Parisi and Terranova, "Heat-Death."

34. Bachelard, *Psychoanalysis of Fire*, 111.

35. Aristotle, *On the Parts of Animals*, 28.

36. Mann, *The Magic Mountain*, 271.

37. Mumford, *Technics and Civilization*, 78, 157, 259.

38. Lévi-Strauss, *The Raw and the Cooked*, 164.

39. Montesquieu, "Of Laws as Relative to the Nature of the Climate," 296.

40. Smith, "'Exceeding Beringia.'"

41. Jennings, *Curing the Colonizers*, 13.

42. Barber, *Modern Architecture and Climate*, 37.

43. Horn, "The Aesthetics of Heat," 2.

44. Horn, "The Aesthetics of Heat," 4.

45. Horn, "The Aesthetics of Heat," 4.

46. Gumbs, *M Archive*, 96; sentences in all lowercase in the original.

47. Gumbs, *M Archive*, 98.

48. Moreover, these workers were never promoted, so over the course of their careers they were not paid the same amount as their white counterparts. Foote, Whatley, and Wright, "Arbitraging a Discriminatory Labor Market."

49. Foote, Whatley, and Wright, "Arbitraging a Discriminatory Labor Market," 494.

50. Abe et al., "Histone Demethylase JMJD1A," 1566; University of Tokyo, "Enduring Cold Temperatures Alters Fat Cell Epigenetics," press release, April 30, 2018, https://www.u-tokyo.ac.jp/focus/en/articles/a_00602.html.

51. Alaimo, *Exposed*, 7.

52. Barber, *Modern Architecture and Climate*; Furuhata, *Climatic Media*.

53. Mukherjee, *Radiant Infrastructures*.

54. While it is true that radiant waves are the medium for any signals communicated, those waves themselves do not require a medium to move through space.

55. Vannini and Taggart, "Making Sense of Domestic Warmth," 66.

56. Soler et al., "Calibration."

57. Heschong, *Thermal Delight in Architecture*, 19.

58. Ong, "Introduction: Environmental Comfort and Beyond," 3.

59. Peterson, "Heat."

60. Barad, *Meeting the Universe Halfway*, 160.

61. Barad, *Meeting the Universe Halfway*, ix.

62. See, for example, the recount of more than 170,000 votes in Palm Beach County, Florida, after voting machines overheated in 2018. Maya Kaufman, "Palm

Beach County's Voting Machines Overheat and Force Recount of More Than 170,000 Votes," *WLRN*, November 14, 2018, https://www.wlrn.org/news/2018-11-14 /palm-beach-countys-voting-machines-overheat-and-force-recount-of-more-than -170-000-votes.

63. In this project, I follow other experimental histories of media, such as Jacqueline Wernimont's *Numbered Lives*, that read mediation speculatively, "making a mess of apparent order in the service of alternative futures." Wernimont, *Numbered Lives*, 3.

64. Deleuze and Guattari, *A Thousand Plateaus*, 410.

65. Furuhata, *Climatic Media*.

66. Marks, *Skin of the Film*, 162.

67. Whyte, "Indigenous Science (Fiction)"; Davis and Todd, "On the Importance of a Date"; and Yusoff, *A Billion Black Anthropocenes or None*.

68. Hartman, *Lose Your Mother*.

69. GilFillan, "The Coldward Course of Progress."

70. Rodriguez-Palacios, Conger, and Cominelli, "Nonmedical Masks in Public."

71. Somini Sengupta, "This Is Inequity at the Boiling Point," *New York Times*, August 7, 2020.

72. Tobías and Molina, "Is Temperature Reducing?"

73. Mandal and Panwar, "Can the Summer Temperatures"; Lin et al., "Containing the Spread."

74. Huntington, "Influenza and the Weather."

75. Matt Rogers, "We Heart Home," Nest.com, April 13, 2107, https://nest.com /blog/2017/04/13/we-heart-home/ (site discontinued).

76. Chun, "On Patterns and Proxies."

77. Watt-Cloutier, *Right to Be Cold*, 230.

Chapter 1: Thermostat

1. Susan S. Lang, "Study Links Warm Offices to Fewer Typing Errors and Higher Productivity," *Cornell Chronicle*, October 19, 2004, http://news.cornell.edu/stories /2004/10/warm-offices-linked-fewer-typing-errors-higher-productivity.

2. Kira Cochrane, "Facebook Staff Feel the Chill in Cold Offices," *Guardian*, March 11, 2013, https://www.theguardian.com/technology/shortcuts/2013/mar/11 /facebook-staff-chill-cold-offices.

3. Kingma and Lichtenbelt, "Female Thermal Demand."

4. Pam Belluck, "Chilly at Work? Office Formula Was Devised for Men," *New York Times*, August 3, 2015.

5. Anthony Lydgate, "Is Your Thermostat Sexist?," *New Yorker*, August 3, 2015, https://www.newyorker.com/tech/annals-of-technology/is-your-thermostat -sexist.

6. Jennifer Horn, "Change the Work Climate," *Stimulant*, August 30, 2016. https://stimulantonline.ca/2016/08/30/change-the-work-climate/.

7. Chang and Kajackaite, "Battle for the Thermostat."

8. Fernández-Galiano, *Fire and Memory*.

9. Barber, *Modern Architecture and Climate*, 26.

10. A notable exception here is *Modern Architecture and Climate*, in which, while tracking the conditioning of architectural environments, Barber observes that the "processes of subject formation begin to pass through thermal interiors, conditioning bodies to experience and engage with specific thermal environments according to physiological norms and expectations, geopolitical and geophysical, mediated by the technical systems of the façade" (73). While I describe here the creation of subject positions through advertisements and technical discourse, Barber shows that this subject's "stasis and position of normativity" (236) are also carefully cultivated through architectural methods.

11. Thermostatic mechanisms include control valves that mix hot and cold water, the human brain's hypothalamus, and, of course, thermostats themselves. In their ideal function, thermostatic mechanisms negotiate hot and cold in order to produce thermal stasis.

12. Ackermann, *Cool Comfort*.

13. Osman, *Modernism's Visible Hand*, 8. Osman documents Andrew Ure's work on the first thermostat in the early 1800s, developed primarily for textile mills but not widely deployed. Thermostatic controls had been designed centuries earlier: in 1666 Cornelius Drebbel sketched an automatic furnace that was regulated by a thermostat. Bedini, "The Role of Automata," 24–42.

14. Adams, *Home Fires*, 94. Literary descriptions of home heating also reflect its gendered orientation. As just one example, Mr. Primrose in *The Vicar of Wakefield* (1766) looks forward to joining his wife and daughters at home, where "smiling looks, a neat hearth, and pleasant fire, were prepared for our reception." Goldsmith, *The Vicar of Wakefield*, 35.

15. Adams, *Home Fires*, 94.

16. Brian Roberts, "Warren S. Johnson and Automatic Controls," Heritage Group Website for the Chartered Institution of Building Service Engineers, accessed November 21, 2018, http://www.hevac-heritage.org/built_environment/pioneers _revisited/johnson.pdf.

17. Osman, *Modernism's Visible Hand*, 32.

18. Atanasoski and Vora, *Surrogate Humanity*, 2.

19. See Cowen, *More Work for Mother*, 96.

20. Aldrich, "An Energy Transition," 2.

21. Even though thermostatic controls emerged in the 1880s, many of these lacked a power source. The spread of electrical power in the twentieth century made the domestic thermostat more effective. "Honeywell T8095 Chronotherm Thermostat," National Museum of American History, accessed November 21, 2018, http://americanhistory.si.edu/collections/search/object/nmah_1392754.

22. Osman, *Modernism's Visible Hand*, 43.

23. Tredgold, *Principles of Warming*, 8.

24. Adams, *Home Fires*.

25. Louis S. Treadwell, "Clean Comfort in the Home," *Scientific American* 135, no. 4 (1926): 268–69.

26. "Family Wealth," *McClure's Magazine*, September 1896, 162; "Even Temperature," *McClure's Magazine*, January 1, 1897, 57; "Shackled at Last," *Cosmopolitan*, October 1896, 725.

27. "Family Wealth," 162.

28. "Shackled at Last," 725.

29. "Was She Incapable?," *Good Housekeeping*, January 1925; found in Minnesota Historical Society Collections, Minneapolis Heat Regulator Co., *1925 Advertising*, 142.E.11.2F.

30. "This Is the Age of Automatic Home Heating," *Ladies' Home Journal*, October 1929; found in Minnesota Historical Society Collections, Minneapolis Heat Regulator Co., *1929 Advertising*, 142.E.11.2F.

31. "What a Relief," *Saturday Evening Post*, October 5, 1929; found in Minnesota Historical Society Collections, Minneapolis Heat Regulator Co., *1929 Advertising*, 142.E.11.2F.

32. "This Is the Age of Automatic Home Heating."

33. Mary Pattison, "Domestic Engineering: The Housekeeping Experiment Station at Colonia, New Jersey," *Scientific American* 106 (April 13, 1912): 330–31, quoted in Osman, *Modernism's Visible Hand*, 37. See Osman for a description of the way that the home, in part through thermostatic technologies, is infused with the values of domestic productivity and efficiency.

34. Osman, *Modernism's Visible Hand*, 21.

35. "New National Campaign"; found in Minnesota Historical Society Collections, Minneapolis Heat Regulator Co., *April 1928 Advertising*, 142.E.11.2F.

36. "Press Stresses Need for Thermostats," *Regulator News*, no. 18 (October 1929), 1.

37. "Little Jimmy," *House Beautiful*, June 1928; found in Minnesota Historical Society Collections, Minneapolis Heat Regulator Co., *June 1928 Advertising*, 142.E.11.2F.

38. "Little Jimmy."

39. "John S. Merlyn," *Blue Island (IL) Sun Standard*, November 29, 1945.

40. "Leveled Heat," *Detroit News*, September 2, 1934; "Chronotherm Control," *New York Times*, October 13, 1935.

41. "Leveled Heat."

42. "Stabilized Heat," *Syracuse Herald*, August 31, 1937.

43. Hutchison, "Journalism and the Perfect Heat Wave."

44. "No! It's *Your* Morning to Turn Up the Thermostat!," *Life*, November 5, 1951, 63.

45. Minnesota Historical Society Collections, Minneapolis Heat Regulator Co., advertising budgets and plans, 1948, 1950-1954.141.A.6.7B, Box 89.

46. Minnesota Historical Society Collections, Minneapolis Heat Regulator Co., advertising budgets and plans, 1948, 1950-1954.141.A.6.7B, Box 89.

47. Minnesota Historical Society Collections, Minneapolis Heat Regulator Co., advertising budgets and plans, 1948, 1950-1954.141.A.6.7B, Box 89.

48. "If You Own One of Those Pre War Thermostats," *New York Times*, September 25, 1949.

49. "Gilding the Thermostat," *Business Week*, November 7, 1953, 94.

50. Minnesota Historical Society Collections, Minneapolis Heat Regulator Co., advertising budgets and plans, 1948, 1950-1954.141.A.6.7B, Box 89.

51. "Would Your Home Reveal 3-Zone Heating?," *New York Times*, November 24, 1935.

52. "Zoned Heat Is Economy to Household," *Hartford (CT) Courant*, January 31, 1943.

53. "Home of Future Seen Provided with Zone Heat," *New York Herald Tribune*, June 6, 1943.

54. "Zoned Heat Is Economy to Household."

55. "Heat Control Can Save Fuel," *Times Building Section*, May 13, 1951.

56. Barber, *Modern Architecture and Climate*.

57. "How Built-in Climate Control Can Give You the Kind of Comfort You've Always Wanted," *Life*, May 11, 1953.

58. Honeywell. "Announcing Honeywell's New Electronic Moduflow," *Life*, September 12, 1955.

59. Honeywell, "Electronic Thermostat," *Life*, October 11, 1954.

60. The Plumbing and Heating Industries Bureau in 1959 described zone heating as the "ultimate" in temperature control; "Zone Heating System," *Hartford Courant*, April 20, 1958.

61. "Heat Control Can Save Fuel."

62. Ralph Mason, "If Your Husband Is a Fresh-Air Fiend," *Baltimore Afro-American*, December 3, 1955.

63. Mason, "If Your Husband Is a Fresh-Air Fiend."

64. Love, "Love and the Thermostat."

65. Lucia Bonaluto, "Thermostat Tyranny Has Only One Cure," *Boston Globe*, February 3, 1974.

66. Edward Cowen, "Nixon Asks Householders to Save Heat by Lowering Thermostat 4 Degrees," *New York Times*, October 10, 1973.

67. Joe DeBona, "Wife Is Burned Up at Joe's Staying Cool," *Hartford (CT) Courant*, November 12, 1973.

68. Menzies, "Thermostat Game Turns Serious in Suburbia."

69. Homes, hotels, and hospitals were exempt. Jack Valenti, head of the Motion Pictures Association of America, expressed disapproval of cooling limits. Originally, the government considered exempting movie theaters.

70. Peter Nulty, "Honeywell Sets Its Thermostat at 'Profit,'" *Fortune*, June 2, 1980, 87.

71. Nulty, "Honeywell Sets Its Thermostat at 'Profit,'" 87.

72. Brezina, "Generating 'Heat' by Turning Down the Thermostat."

73. Geist, "Fiddling with the Thermostat."

74. Peter Tonge, "Heating a House with Rolled Newspapers," *Christian Science Monitor*, December 17, 1980.

75. Joseph N. Bell, "Cold-Hot Relationships Require a Degree of Patience," *Los Angeles Times*, February 17, 1990.

76. Bell, "Cold-Hot Relationships."

77. *United States ex rel. Davis v. Gramley*, Case No. 98 C 1452 (N.D. Ill. Mar. 26, 2008).

78. *Boggess v. Roper*, No. 3:04cv92 (W.D.N.C. Sep. 1, 2006); *Hale v. Village of Madison*, 493 F.Suppl.2d 928 (N.D. Ohio 2007); *Howard v. Burns Bros.*, 149 F.3d 835 (8th Cir. 1998).

79. *Stanina v. Blandin Paper Co.*, Civil No. 04-3804 (JRT/FLN) (D. Minn. Jul. 26, 2006).

80. *Doe v. Forrest*, 176 Vt. 476, 2004 Vt. 37, 853 A.2d 48 (Vt. 2004).

81. "Wife Sues, So Husband Blasts Home," *Washington Post and Times Herald*, March 8, 1959.

82. Nellie Bowles, "Thermostats, Locks, and Lights: Digital Tools of Domestic Abuse," *New York Times*, June 23, 2018.

83. Comstock, "A Long, Strong Pull Together."

84. Comstock, "A Long, Strong Pull Together."

85. Comstock, "A Long, Strong Pull Together."

86. Cooper, *Air-Conditioning America*, 2

87. See, for example, Hough and Sedgwick, *Human Mechanism*.

88. "Report of New York State Commission on Ventilation."

89. "Report of New York State Commission on Ventilation."

90. The laboratory was set up following the conclusion of the report in 1917 but before its results were made public in 1923. Cooper, *Air-Conditioning America*, 70.

91. They also collaborated with other programs established at the same time, including the Harvard School of Public Health's industrial hygiene program.

92. Allen, "Modus Operandi."

93. This wasn't the first "comfort chart"; John Wilkes Sheppard of the Teachers' Normal College in Chicago in 1916 also developed a comfort chart that defined human comfort in terms of temperature and humidity. This chart was also in part an adaptation of 1911 chart presented by Willis Carrier, dubbed the "father of air-conditioning." Cooper, *Air-Conditioning America*, 70.

94. Houghten and Yagloglou, "Determination of the Comfort Zone"; Houghten and Yagloglou, "Determining Equal Comfort Lines."

95. Houghten and Yagloglou, "Determination of the Comfort Zone," 365.

96. Houghten and Yagloglou, "Determination of the Comfort Zone," 366.

97. Barber, *Modern Architecture and Climate*, 200.

98. De Dear and Brager, "Developing an Adaptive Model," 1.

99. ISO 7243, ISO 7933, and ISO/TR 11079 deal with extreme environmental conditions.

100. American Society of Heating, Refrigerating and Air-Conditioning Engineers (ASHRAE), Standard 55-2017.

101. Heijs and Stringer, "Research on Residential Thermal Comfort," 237.

102. Van Hoof, "Forty Years of Fanger's Model"; McIntyre, "Evaluation of Thermal Discomfort."

103. ASHRAE, Standard 55-2013, 4.

104. Kempton and Lutzehiser, introduction, 172.

105. Parsons, introduction, xiii.

106. Parsons, introduction, xiii.

107. Agbemabiese, Berko, and du Pont, "Air Conditioning in the Tropics."

108. Agbemabiese, Berko, and du Pont, "Air Conditioning in the Tropics."

109. De Dear and Brager, "Developing an Adaptive Model," 1.

110. De Dear and Brager, "Developing an Adaptive Model," 2.

111. De Dear and Brager, "Developing an Adaptive Model," 2.

112. De Dear and Brager, "Developing an Adaptive Model," 2.

113. Humphreys, "Thermal Comfort Temperatures," 5.

114. Humphreys, "Thermal Comfort Temperatures," 13.

115. De Dear and Brager, "Developing an Adaptive Model," 2.

116. Parsons, *Human Thermal Environments*, xxi.

117. Agbemabiese, Berko, and du Pont, "Air Conditioning in the Tropics."

118. Rohles, "Temperature and Temperament," 14.

119. Parsons, introduction, xiv.

120. Van Hoof, "Forty Years of Fanger's Model."

121. Farhad Manjoo, "Home Thermostats, Wallflowers No More," *New York Times*, December 7, 2011.

122. Edward C. Baig, "Review: Newfangled Nest Thermostat Is Hot," *USA Today*, October 26, 2011; Matt Burns, "Honeywell vs Nest: When the Establishment Sues Silicon Valley," *TechCrunch*, February 7, 2012, https://techcrunch.com/2012/02/07/honeywell-vs-nest-when-the-establishment-sues-silicon-valley; David Pogue, "A Thermostat That's Clever, Not Clunky," *New York Times*, December 1, 2011.

123. Parks Associates, "Eight Percent of U.S. Broadband Households Own a Smart Thermostat," *MarketWired*, April 14, 2015, http://www.marketwired.com/press-release/parks-associates-eight-percent-of-us-broadband-households-own-a-smart-thermostat-2009535.htm.

124. Kirsten Korosec, "Honeywell Sues Nest Labs, Best Buy over Smart Thermostat Tech," *ZDNet*, February 6, 2012, https://www.zdnet.com/article/honeywell-sues-nest-labs-best-buy-over-smart-thermostat-tech.

125. "Honeywell vs. Nest: The Battle for the Smart Thermostat," *Predicting Our Future Podcast*, January 11, 2018, https://www.predictingourfuture.com/12-honeywell-vs-nest-the-battle-for-the-smart-thermostat.

126. Jessica Salter, "Tony Fadell, Father of the iPod, iPhone and Nest, on Why He Is Worth $3.2bn to Google," *Telegraph*, November 14, 2014.

127. Andrew Weinreich, "The Future of the Smart Home: Honeywell vs. Nest: The Battle for the Smart Thermostat," *Forbes*, January 11, 2018.

128. Furuhata, *Climatic Media*.

129. Nest, "Nest Learning Thermostat," accessed December 4, 2018, https://nest.com/thermostats/nest-learning-thermostat/overview.

130. Ecobee, "Donate Your Data," accessed December 4, 2018, https://www.ecobee.com/donateyourdata.

131. Salter, "Tony Fadell."

132. Nest, "Nest Learning Thermostat."

133. Nest, "Nest Learning Thermostat."

134. Schüll, "Data for Life," 13.

135. Carrier Corporation, "Thermostats," accessed December 4, 2018, https://www.carrier.com/residential/en/us/products/thermostats.

136. Kiernan Salmon, "Using TherMOOstat to Find Comfort in Your Lecture Hall," UC Davis Energy Conservation Office (blog), March 13, 2017, https://facilitiesmanagement.sf.ucdavis.edu/blog/crowdsourcing-keep-classrooms-comfortable.

137. Other technologies enable their users or operators to control temperature through the manipulation of building materials rather than central heating or cooling systems. In buildings with View's Dynamic Glass, for example, the glass's tint can be changed to increase or decrease the amount of solar radiation in the room—and, by extension, to raise or lower the temperature.

138. Cooper, Air-Conditioning America, 157.

139. Cooper, Air-Conditioning America, 158.

140. Cooper, Air-Conditioning America, 158.

141. Aaron Freedman, "Freezing Workers of the World, Unite!," Jacobin, July 9, 2019.

142. Steven Dawson-Haggerty, "Funding Announcement," Comfy News and Events (blog), November 7, 2013, https://www.comfyapp.com/blog/funding-announcement.

143. Günel, Spaceship in the Desert, 120.

144. Nick Bilton, "Nest Thermostat Glitch Leaves Users in the Cold," New York Times, January 13, 2016.

145. Susan-Elizabeth Littlefield, "Nest Thermostat Die amid Arctic Blast? Here's a Quick Fix," CBS Minnesota, December 26, 2017, https://minnesota.cbslocal.com/2017/12/26/nest-cold-weather-glitch.

146. @targetsalesbin, "Below freezing outside and the device failed, allowing our home inside to drop to 50°F this morning," December 27, 2017, Twitter. Original source no longer available.

147. Chris Welch, "Nest Will Provide 1 Million Thermostats to Low-Income Homes as Part of New Power Project," Verge, April 19, 2018, https://www.theverge.com/2018/4/19/17256336/nest-power-project-thermostat-low-income-families; Nest Power Project, "Everyone Deserves Light in the Dark and Heat in the Cold," accessed December 4, 2018, https://nestpowerproject.withgoogle.com.

148. Jeff St. John, "Inside Nest's 50,000-Home Virtual Power Plant for Southern California Edison," GreenTechMedia, accessed December 4, 2018, https://www.greentechmedia.com/articles/read/inside-nests-50000-home-virtual-power-plant-for-southern-california-edison.

149. St. John, "Inside Nest's 50,000-Home Virtual Power Plant."

150. Miller, Internet of Things, 98.

151. Chuck Martin, "Smart Heating App Exposed as 'Burglar's Dream,'" Media-Post, August 20, 2015.

152. Lorenzo Franceschi-Bicchierai, "Hackers Make the First-Ever Ransomware for Smart Thermostats," *Vice*, August 7, 2016, https://motherboard.vice.com/en_us/article/aekj9j/internet-of-things-ransomware-smart-thermostat.

153. Nest, "Terms of Service," accessed December 4, 2018, https://nest.com/legal/terms-of-service.

154. Barber, *Modern Architecture and Climate*, 262.

155. Furuhata, *Climatic Media*.

Chapter 2: Coldsploitation

1. Bradshaw, *Building Environment*, 4; emphasis added.

2. Bachelard, *Psychoanalysis of Fire*, 38.

3. Heschong, *Thermal Delight*.

4. Ong, "Warming Up to Heat."

5. Candido and de Dear, "From Thermal Boredom."

6. Basile, *Cool*, inside cover.

7. Shachtman, *Absolute Zero and the Conquest of Cold*.

8. DeBow, "Ice."

9. Thévenot, *History of Refrigeration*, 130–31.

10. Rees, *Refrigeration Nation*, 8, 10.

11. Hobart, "Cooling the Tropics," n.p.

12. Hobart, "Cooling the Tropics."

13. See also, for example, Herold, "Ice in the Tropics."

14. "Ice in the Tropics," *Ice and Refrigeration* 3, no. 6 (1892): 436.

15. As the editors of *After Ice* observe, water's phase transitions "offer modes of colonial concealment" that sediment settler practices in frozen ecologies. Ruiz, Schönach, and Shields, "Introduction: Experiencing after Ice."

16. As a critical example, Rebecca J. H. Woods documents the interconnection of the frozen meat trade with the development of new breeds of sheep for mutton exports. Woods, *Herds Shot Round the World*. See also Osman, *Modernism's Visible Hand*.

17. Blum, *News at the Ends of the Earth*, 38. Well before these expeditions, as Christopher P. Heuer argues, the North offered a fundamentally "different kind of terra incognita for the Renaissance imagination," one that "powerfully challenged older understandings of image-making." Heuer, *Into the White*, 10.

18. J. R. Smith, "Ice Core Coloniality."

19. Ackermann, *Cool Comfort*, 19.

20. Ruiz, Schönach, and Shields, "Introduction: Experiencing after Ice."

21. Marks, *Skin of the Film*, xii.

22. Furuhata, *Climatic Media*.

23. Howe, "On Cryohuman Relations."

24. Thompson and Bordwell, *Film History*, 144.

25. Hansen, *Feed-Forward*.

26. Sloterdijk, *Terror from the Air*, 18.

27. Sloterdijk, *Terror from the Air*, 25.

28. Virilio, *War and Cinema*, 1.

29. "Snow Makes War White along Somme," *Los Angeles Times*, January 16, 1917; "Armies Fighting in Varied Climates," *New York Times*, November 11, 1914.

30. "Winter and the War," *Boston Daily Globe*, February 28, 1915.

31. "Storm Cripples City," *New York Tribune*, December 15, 1917.

32. "16 Months in Antarctic Cold Rivals Stories of Heroism on War Front," *Washington Post*, December 3, 1916.

33. James B. Wharton, "With Amundsen in the Arctic," *Atlanta Constitution*, July 12, 1925.

34. MacKenzie and Stenport, *Films on Ice*, 5.

35. "The Iron Trail," *Exhibitor's Trade Review*, November 12, 1921, 1701.

36. Schaefer, *"Bold! Daring! Shocking! True!"* There is some evidence that "stories of the snows" appealed to the states' rights market, in which classic exploitation was circulated and which was often oriented toward the affective rather than intellectual appeal.

37. "'Cold Feet' Surely Will Warm Up Any Kind of Old House," *Exhibitor's Trade Review*, April 15, 1922, 1402.

38. "Chechahcos Was Really Made up in Alaska," *Star-Phoenix Saskatoon*, October 3, 1925.

39. "Two Releases from Paramount Week of June 4," *Exhibitor's Trade Review*, June 10, 1922, 83.

40. "The Heart of the North," *Exhibitor's Trade Review*, September 10, 1921, 1054.

41. "Shadows of Conscience," *Exhibitor's Trade Review*, October 29, 1921, 1557.

42. "Astounding Climax Caps Griffith's Latest Screen Sensation," *Film Daily*, September 12, 1920.

43. "The Young Diana," *Exhibitor's Trade Review*, September 16, 1922, 1077.

44. "Antarctic Pictures of Mawson Journey Unusual in Interest," *Exhibitor's Trade Review*, June 24, 1922.

45. "Striking the Keynote," *Exhibitor's Trade Review*, September 9, 1922, 987.

46. "Capitol, New York, Books 'Nanook,'" *Exhibitor's Trade Review*, June 3, 1922, 16.

47. "'Nanook' Wins Favor in Dallas," *Exhibitor's Trade Review*, July 8, 1922, 331.

48. Ferguson, "Panting in the Dark."

49. E. E. Barrett, "Snow Men and Women," *Pictures and the Picturegoer*, December 1, 1924, 40–41.

50. "Snow Stuff," *Picturegoer* 2, no. 12 (1921): 10–11.

51. Daisy Dean, "News and Notes from Movieland," *Charlotte (NC) News*, February 6, 1922.

52. Marjorie Charles Driscoll, "Freeze-Outs De Luxe," *Picture-Play Magazine*, September 1921, 29.

53. "Snow Stuff."

54. "Behind the Screen," *Pictures*, May 7, 1921, 447.

55. Mollie Gray, "Women's Page," *Variety*, July 18, 1928, 41.

56. Vaughan, *Hollywood's Dirtiest Secret*, 106.

57. Gaines, "From Elephants to Lux Soap."

58. "Striking the Keynote," *Exhibitor's Trade Review*, September 9, 1922, 987.

59. "Exploitation Ideas," *Exhibitor's Trade Review*, June 14, 1924, 35.

60. "Atmospheric Setting," *Exhibitor's Trade Review*, March 11, 1922, 1055.

61. "Iron Trail."

62. "Prologuing with Modern Effects," *Exhibitor's Trade Review*, October 21, 1921, 1533.

63. "The Light Artist's Palette," *Motion Picture News*, February 1, 1930, 78.

64. "Advertising the Picture Theater," *Exhibitor's Trade Review*, September 2, 1922, 949.

65. "Frozen Pictures," *Exhibitor's Trade Review*, August 19, 1922, 830.

66. "Money-Making Suggestions for Ambitious Merchants," *Talking Machine World*, November 15, 1926, 47.

67. Simmons, "Settler Atmospherics."

68. Charles Reisner, "Nanook of the Girl," *Exhibitors Herald*, December 25, 1926, 26.

69. Blum, *News at the Ends of the Earth*, 38.

70. Smith, "'Exceeding Beringia.'"

71. "'The Storm' as Campaign Inspiration," *Exhibitor's Trade Review*, September 23, 1922, 1133.

72. Grant Heth, "Screen Must Be Built to Confirm," *Exhibitor's Trade Review*, October 15, 1921, 1406.

73. Monte Sohn, "Baker on Rebuilding Patronage," *Exhibitor's Trade Review*, November 19, 1921, 1711.

74. "Striking the Keynote," *Exhibitor's Trade Review*, September 9, 1922, 987.

75. "The Voice of the Box Office," *Exhibitor's Trade Review*, September 17, 1921, 1099.

76. Sargent, *Picture Theater Advertising*, 274.

77. Guy P. Leavitt, "Is the Airdome Doomed in the Middle West?," *Motion Picture News*, July 29, 1916, 624.

78. "*Out of the Silent North*," *Exhibitor's Trade Review*, June 24, 1922, 241.

79. "The Voice of the Box Office," *Exhibitor's Trade Review*, September 17, 1921, 1100.

80. Leavitt, "Is the Airdome Doomed?"

81. Tom Kennedy, "Making the Theater Pay," *Exhibitor's Trade Review*, September 16, 1922, 1063.

82. Walter F. Eberhardt, "Hot Stuff—Now You Get It," *Exhibitor's Trade Review*, June 24, 1922, 221.

83. Eberhardt, "Hot Stuff."

84. Eberhardt, "Hot Stuff."

85. "Trade Notes," *Motion Picture World*, August, 10 1907, 358.

86. Eberhardt, "Hot Stuff."

87. "Only Cooling Hymns," *Austin (TX) Statesman*, August 14, 1910.

88. "Vorbach Bros. Help to Lower the Temperature," *Talking Machine World*, September 15, 1923, 74.

89. Gunning, "Before Documentary," 15.

90. Eberhardt, "Hot Stuff."

91. Eberhardt, "Hot Stuff."

92. Sargent, *Picture Theater Advertising*, 276.

93. "Emergency Action," *Exhibitor's Trade Review*, July 8, 1922, 357.

94. Audrey Amidon, "The Aviator, the Explorer, and the Radio Man: The 1925 MacMillan Arctic Expedition," *Unwritten Record* (blog), September 30, 2015, https://unwritten-record.blogs.archives.gov/2015/09/30/the-aviator-the-explorer-and-the-radio-man-the-1925-macmillan-arctic-expedition/.

95. Vance, "Broadcasting from the Polar Snows," *Radio Progress*, September 15, 1925, 25–28.

96. "Zenith at the North Pole," *Radio Age* 1, 1924, p. 52

97. J. L. Wilkie, "The 'Farthest North' Broadcasting Station," *Radio World*, February 17, 1923, 7.

98. "Byrd Expedition to Use Kolster Products," *Talking Machine World*, August 1928, 127.

99. "Where Portability Counts," *Moving Picture World*, April 21, 1923, 878.

100. "Tuesday Programs," *Broadcast Weekly*, January 27, 1929, 38.

101. Taylor, *The Sounds of Capitalism*, 33.

102. "A Tropical-Hot Orchestra," *Tune In*, November 1945, 10.

103. "March 24," *Popular Radio*, April 1927, 372.

104. "Profit Winning Sales Wrinkles," *Talking Machine World*, May 1928, 14.

105. "Exploiting Your Cooling System," *Exhibitor's Trade Review*, October 29, 1921, 1534.

106. Sargent, *Picture Theater Advertising*, 275.

107. "In the Frozen North," *Moving Picture World*, June 3, 1911, 1279.

108. "Clean, Cool, Fresh Air Will Boost Your Box Office Receipts," *Exhibitor's Trade Review*, July 8, 1922, 375.

109. Lovecraft, "Cool Air."

110. Lovecraft, "Cool Air."

111. Lovecraft, "At the Mountains of Madness," 17.

112. Lindsay, "Air Conditioning as Applied."

113. Simonds and Polderman, "Air Conditioning in Film Laboratories," 605.

114. George Schutz, "He Did Something about the Weather," *Motion Picture Herald*, March 6, 1937, 14.

115. Cooper, *Air-Conditioning America*, 80.

116. Simonds and Polderman, "Air Conditioning in Film Laboratories."

117. Buensod, "Theater Air Conditioning," 539.

118. For more on neutralized theater architecture, see Szczepaniak-Gillece, *The Optical Vacuum*.

119. Baker, "Air Cleaning and Conditioning," 4.

120. Lindsay, "Air Conditioning as Applied," 336.

121. J. T. Knight, "What Makes the Most Efficient Theatre Air-Conditioning System?," *Motion Picture Herald*, March 6, 1937, 16.

122. Ackermann, *Cool Comfort*, 68.

123. Lindsay, "Air Conditioning as Applied," 334.

124. Lindsay, "Air Conditioning as Applied," 335.

125. Knight, "What Makes the Most Efficient?," 16, 17.

126. Lindsay, "Air Conditioning as Applied," 335.

127. Knight, "What Makes the Most Efficient?," 17.

128. Gordon H. Simmons, "Why Not a Spirit of More Fraternity amongst our Frantic Idea Men?," *BoxOffice*, January 4, 1941, 57.

129. Gordon H. Simmons, "Proper Control Is the Crux of an Air Conditioning Plant," *BoxOffice*, March 29, 1941, 68.

130. Simmons, "Why Not a Spirit?," 56.

131. Chesapeake and Ohio Railroad, "Air-Conditioning on the Air," *Broadcasting*, October 15, 1932, 18.

132. "Philco Adds Air-Conditioning," *Radio Today*, June 1938, 70.

133. "Philco Adds Air-Conditioning," 68.

134. Gaines, "From Elephants to Lux Soap," 32.

135. World Tourism Organization, "Climate Change and Tourism," accessed December 17, 2017, http://sdt.unwto.org/en/content/climate-change-tourism.

136. hooks, "Eating the Other," 21.

137. Interestingly, as Rafico Ruiz points out, this rationale also surrounded iceberg-harvesting efforts in the 1970s: heat differential was imagined as a critical way to generate energy. Ruiz, "Phase State Earth." See also Roberts, "Icebergs as a Heat Sink."

138. Beretta et al., "Thermoelectrics," 213.

139. Beretta et al., "Thermoelectrics," 214.

140. Drew Prindle, "Warm Up or Cool Down with the Press of a Button on the Wrist-Worn Embr," *Digital Trends*, June 2, 2018, https://www.digitaltrends.com/cool-tech/embr-wave-review.

141. Embr, "Embr Wave Bracelet," accessed December 4, 2018, https://embrlabs.com/embr-wave.

142. Kishore et al., "Ultra-High Performance."

143. Wang et al., "The Effect of a Low-Energy Wearable Thermal Device on Human Comfort," 2.

144. Humphreys, "Thermal Comfort Temperatures," 10.

145. Cabanac, "Sensory Pleasure."

Chapter 3: Sweatbox

Portions of this chapter were previously published as "Thermal Violence: Heat Rays, Sweatboxes and the Politics of Exposure," *Culture Machine* 17 (2018), https://culturemachine.net/vol-17-thermal-objects/thermal-violence/.

1. Razack, *Dying from Improvement*, 172.

2. Sákéj Henderson, director of the Native Law Centre, quoted in DeNeen L. Brown, "Left for Dead in a Saskatchewan Winter," *Washington Post*, November 22, 2003.

3. Purdon, "On the Use of Ice," 379.

4. Showalter, "Hysteria, Feminism, and Gender," 286.

5. Purdon, "On the Use of Ice," 379.

6. "Influence of the Cerebellum on the Genitals," 168.

7. *Jane Doe et al. v. Jeh Johnson et al.*, No. 16-cv-00750 W (BLM) (S.D. Cal. Jul. 3, 2018).

8. Aura Bogado, "Inside the Immigration 'Icebox,'" *Colorlines*, July 11, 2014.

9. Cantor, *Hieleras (Iceboxes)*.

10. *Unknown Parties v. Nielsen*, No. CV-15-00250-TUC-DCB (D. Ariz. Mar. 15, 2019); *Doe v. Johnson*, No. 16-cv-00750 W (BLM) (S.D. Cal. Jul. 3, 2018).

11. Andrew Gumbel, "'They Were Laughing at Us': Immigrants Tell of Cruelty, Illness and Filth in US Detention," *Guardian*, September 12, 2018.

12. De León, *Land of Open Graves*, 213.

13. Decades before that, insurrections on Soviet prison ships were suppressed with fire hoses and prisoners' bodies were frozen together. See Bollinger, *Stalin's Slave Ships*, 102.

14. Krause, *Battle for Homestead*.

15. Michael McLaughlin, "North Dakota Governor Orders Pipeline Protesters to Evacuate," *Huffington Post*, November 28, 2016, https://www .huffingtonpost.com/entry/army-corps-engineers-dakota-access-pipeline-evict_us _583cb3e1e4b04b66c01b832b.

16. Nixon, *Slow Violence*, 2.

17. Radin and Kowal, "The Politics of Low Temperature," 5.

18. Gilmore, *Golden Gulag*, 28.

19. Tuck, "Suspending Damage."

20. Tuck, "Suspending Damage," 417.

21. Sifakis, "Sweatbox," 251.

22. Thompson, "Circuits of Containment."

23. Interview with Prince Smith, in Federal Writers' Project, *Slave Narratives*, vol. 14, *South Carolina*, part 4, 117–18, https://www.loc.gov/item/mesn144.

24. Interview with Levi D. Shelby Jr., in Federal Writers' Project, *Slave Narratives*, vol. 1, *Alabama*, 159–60, https://www.loc.gov/item/mesn010/.

25. Interview with Jessie Sparrow, in Federal Writers' Project, *Slave Narratives*, vol. 14, *South Carolina*, part 2, 139, https://www.loc.gov/item/mesn142/; Interview with Gable Locklier, in Federal Writers' Project, *Slave Narratives*, vol. 14, *South Carolina*, part 3, 113, https://www.loc.gov/item/mesn143/.

26. Washington, *Medical Apartheid*, 29.

27. Davis, "Reflections on the Black Woman's Role," 85.

28. Interview with Janie Scott, in Federal Writers' Project, *Slave Narratives*, vol. 1, *Alabama*, 339, https://www.loc.gov/item/mesn010/.

29. Chol-hwan and Rigoulot, *Aquariums of Pyongyang*, 96.

30. Massachusetts General Court, *Investigation into the Management and Discipline*, 145.

31. "The Inquisition Revived," *Wisconsin Daily Patriot* (Madison), November 24, 1860.

32. "Torture on Board the Pawnee," *Trenton (NJ) State Gazette*, June 7, 1860.

33. W. Sherman, P.A. Surgeon, "Special Dispatch to the New-York Times," *New York Times*, October 20, 1858.

34. At the Mansfield Reformatory in Ohio, "one African-American inmate reports of being disciplined by being placed in a 'sweatbox,' a special type of torture that white prisoners escaped." "Mansfield Reformatory," accessed November 1, 2019, http://www.deadohio.com/mansfieldreformatory.htm (site discontinued).

35. "The Sweat Box State," *Chicago Defender*, September 24, 1927.

36. Lichtenstein, *Twice the Work*, 183.

37. Welch, "Chain Gangs," 108–9.

38. Calls for sweatbox-like practices as a "humane" alternative to "direct" punishment can be dated even to the mid-nineteenth century, when a visitor to the Auburn, New York, prison was assured that the infliction of stripes had "been wholly laid aside" and in lieu of this "they have adopted the *cold shower bath*, and find it, thus far, fully adequate to all punitive purposes." Packard, *Memorandum*, 4.

39. Perkinson, "Hell Exploded," 64.

40. Perkinson, "Hell Exploded," 64.

41. Florida Department of Corrections, "1921," accessed January 10, 2018, http://www.dc.state.fl.us/oth/timeline/1921.html (site discontinued).

42. Esposito and Wood, *Prison Slavery*, 132.

43. "Sick Convict Put in 'Sweatbox'; Find Him Dead," *Chicago Daily Tribune*, September 7, 1927.

44. Giovacchini, *Hollywood Modernism*, 55.

45. The imprisonment of white men in sweatboxes would continue to pervade the white imagination in popular films such as *The Bridge on the River Kwai* (1957) and *Cool Hand Luke* (1967).

46. "Florida Convicts Strike at Sweat-Box Torture," *Washington Post*, October 28, 1932.

47. "'Sweatbox Case' Ruling Is Upheld," *Atlanta Constitution*, September 13, 1942.

48. Rejali, *Torture and Democracy*, 74.

49. *Suhail Najim Abdullah al Ahimari et al. v. Caci International, Inc., et al.*, 951 F. Supp.2d 857 (E.D. Va. 2013).

50. With the coming of electrical power, another technological form was enlisted into thermal manipulation: electricity. In the absence of architectures that could redirect the heat of the sun or boilers, bright electric lights could be used to "sweat" suspects. Rejali, *Torture and Democracy*, 128.

51. Waldrep, *Jury Discrimination*, 213.

52. There has been much research on the historical transformation of these knowledges. See, for example, Hoberman, *Black and Blue*, on the doctrine of Black "hardiness"; Willoughby, "'His Native, Hot Country,'" on physicians' evolving understanding of the relationship between racial difference and climate in the United States; and Jennings, *Curing the Colonizers*, on how the acclimatization debates shaped the politics of French colonial settlement.

53. Washington, *Medical Apartheid*, 40

54. Hartman, *Lose Your Mother*.

55. Florida Department of Corrections, "Union Correctional Institution," accessed January 10, 2018, http://www.dc.state.fl.us/facilities/region2/213.html (site discontinued).

56. Florida Department of Corrections, "1956," accessed January 10, 2018, http://www.dc.state.fl.us/oth/timeline/1956–1961.html (site discontinued).

57. William Booth, "Without the 'Rock' Florida Inmates Get the Hard Place," *Washington Post*, October 30, 1994.

58. David Martin, "A Day of My Life," American Prison Writing Archive at Hamilton College, April 13, 2015, https://apw.dhinitiative.org/islandora/object/apw%3A12343298?solr_nav%5Bid%5D=2291b291c10c614bd1a8&solr_nav%5Bpage%5D=0&solr_nav%5Boffset%5D=0.

59. Florida Department of Health, "1998 Extreme Heat," accessed April 5, 2021, http://www.floridahealth.gov/environmental-health/climate-and-health/_documents/extreme-heat-factsheet.pdf.

60. Booth, "Without the 'Rock.'"

61. *Chandler v. Crosby*, 379 F.3d 1278 (11th Cir. 2004).

62. Vassallo, "Report on the Risks"; Winter and Hanlon, "Parchman Farm Blues."

63. American Civil Liberties Union, "Court Finds 'No Excuse' for Deplorable Conditions on Mississippi's Death Row, Orders Immediate Remedies," May 21, 2003, https://www.aclu.org/news/court-finds-no-excuse-deplorable-conditions-mississippis-death-row-orders-immediate-remedies.

64. American Civil Liberties Union, "ACLU Returns to Court on Behalf of Women at Medical Risk in Sweltering Baltimore City Jail," August 6, 2003, https://www.aclu.org/news/aclu-returns-court-behalf-women-medical-risk-sweltering-baltimore-city-jail.

65. American Civil Liberties Union, "Dangerous Threat of Heat-Related Illness Prompts Appeals Court to Affirm Order to Cool Supermax Prison," July 2, 2004, https://www.aclu.org/news/dangerous-threat-heat-related-illness-prompts-appeals-court-affirm-order-cool-supermax-prison.

66. American Civil Liberties Union, "Dangerous Threat."

67. Darius Rejali, "Ice Water and Sweatboxes," *Slate*, March 17, 2009, http://www.slate.com/articles/news_and_politics/jurisprudence/2009/03/ice_water_and_sweatboxes.html.

68. Jerry Iannelli, "Rundle Won't Charge Prison Guards Who Allegedly Boiled Schizophrenic Black Man to Death," *Miami New Times*, March 17, 2017.

69. Central Intelligence Agency, *Counterterrorism Detention*, 75.

70. The Office of Medical Services' "Guidelines on Medical and Psychological Support to Detainee Interrogations" provides a "safe temperature range" for interrogation. Central Intelligence Agency, *Counterterrorism Detention*.

71. Hensley et al., "50 Years of Computer Simulation," 1.

72. Thomas Littek, "A Concise History of the Prisoners' Rights Movement: An Epic Struggle for Human Dignity," American Prison Writing Archive at Hamilton College, April 13, 2015, https://apw.dhinitiative.org/islandora/search/?f%5B0%5D=mods_name_personal_author_namePart_s%3A%22Littek%2C%20Thomas%22.

73. Ramon Pequero, "Home Sick: I Can't Wait to Go Home," American Prison Writing Archive at Hamilton College, July 3, 2018, https://apw.dhinitiative.org

/islandora/object/apw%3A12347089?solr_nav%5Bid%5D=b2733b22ef53c43d9a82&solr_nav%5Bpage%5D=0&solr_nav%5Boffset%5D=0.

74. Prison Vitality, "Caged View—April 10, 2015," American Prison Writing Archive at Hamilton College, April 10, 2015, https://apw.dhinitiative.org/islandora/search/catch_all_fields_mt%3A%28%22caged%20view%22%29.

75. Muti-Ajamu Osagboro, "Torture through a 'Child's' Eyes: The Hole at Hellzdale Prison—A Report," American Prison Writing Project at Hamilton College, accessed April 5, 2021, https://apw.dhinitiative.org/islandora/object/apw%3A12349019?solr_nav%5Bid%5D=3a3e5e4c7370d2fb52b1&solr_nav%5Bpage%5D=0&solr_nav%5Boffset%5D=4.

76. Shakur, "Women in Prison," 8.

77. Dillon, "Possessed by Death," 114.

78. Equiano, "Interesting Narrative," 76.

79. Parr, *Sensing Changes*, 4, 8.

80. Peterson, "Sensory Attunements," 93.

81. Incarcerated Workers Organizing Committee, "Prison Strike 2018," accessed September 1, 2018, https://incarceratedworkers.org/campaigns/prison-strike-2018.

82. Leigh Phillips, "In Defense of Air-Conditioning," *Jacobin*, August 30, 2018, https://jacobinmag.com/2018/08/air-conditioning-climate-change-energy-pollution.

83. And in this case, raising money for air-conditioning units doesn't necessarily help because the schools' electrical systems can't handle the increased power draw anyway. Mike Newall, "How Philly's Overheated Schools Show Disregard for Our Students," *Philadelphia Inquirer*, September 5, 2018.

84. Sharpe, *In the Wake*, 106.

85. See, for example, Karen Heller, "I Don't Need Air Conditioning, and Neither Do You," *Washington Post*, August 18, 2016.

86. Vassallo, "Report on the Risks."

87. Alaimo, *Bodily Natures*.

88. Alaimo, *Exposed*, 94.

89. Mukherjee, *Radiant Infrastructures*, 142.

90. Yusoff, *Billion Black Anthropocenes*.

Chapter 4: Heat Ray

Portions of this chapter were previously published as "Thermal Violence: Heat Rays, Sweatboxes and the Politics of Exposure," *Culture Machine* 17 (2019), https://culturemachine.net/vol-17-thermal-objects/thermal-violence/.

1. LeVine, *Active Denial System*, 6.

2. Noah Shachtman, "Pain Ray: Don't Hold Your Breath," *Wired*, December 10, 2007.

3. Margaret Winter and Peter J. Eliasberg, "Don't Let the Military's Deadly 'Pain Ray' Machine Invade the L.A. County Jail," *Speak Freely* (blog), August 26, 2010, https://www.aclu.org/blog/national-security/dont-let-militarys-deadly-pain-ray-machine-invade-la-county-jail.

4. Wells, *War of the Worlds*, 45.

5. Mukherjee, *Radiant Infrastructures*.

6. Tawil-Souri, "Spectrum."

7. Rossi, "High-Performance Sportswear."

8. Lisa Zyga, "Super-Insulated Clothing Could Eliminate Need for Indoor Heating," Phys.org, January 8, 2015, https://phys.org/news/2015-01-super-insulated -indoor.html.

9. A notable exception has been the examination of how textiles contend with the elements, including wind, rain, heat, and cold. See Ruiz, "Grenfell Cloth."

10. "Bell's Photophone," *Scientific American*, January 1, 1881, 1.

11. Prescott, *Bell's Electric Speaking Telephone*, 313.

12. Bruce, *Alexander Graham Bell*, 337.

13. Seemann et al., *Agrometeorology*, 28; US Department of Commerce, National Bureau of Standards, "Letter Circular 686: Photoelectric Cells; Selenium Cells; Thermopiles," NBS *Letter Circulars*, April 4, 1942.

14. Robert C. Brasted, "Selenium," *Encyclopaedia Britannica*, online ed., accessed August 28, 2019, https://www.britannica.com/science/selenium.

15. W. Smith, "Action of Light on Selenium."

16. Bell, "Selenium and the Photophone."

17. Bell, "Upon the Production of Sound," 41–42.

18. Wells, *War of the Worlds*, 45.

19. Mims, "First Century of Lightwave Communications," 16.

20. Bate and Morrison, "Some Aspects," 10.

21. Case, "Infra Red Telegraphy and Telephony," 406.

22. John Stuart, "Telegraphing with Invisible Rays of Heat," *Popular Science Monthly*, September 1920, 51.

23. "Infrared-Ray Telegraph Has Interesting Equipment," *Popular Mechanics*, May 1920, 649.

24. Rankine, "Transmission of Speech by Light," 744.

25. J. L. Baird, "Television in 1932," from BBC Annual Report, 1933, *Baird Television: A Website about Television History*, accessed September 22, 2019, http://www .bairdtelevision.com/1932.html.

26. M. H. I. Baird, "Baird in America," *Baird Television: A Website about Television History*, accessed September 22, 2019, http://www.bairdtelevision.com/America.html.

27. M. H. I. Baird, "Baird in America."

28. Burns, *John Logie Baird*, 112.

29. Interestingly, television was not Baird's first thermal invention. Before these experiments, Baird—who was known for having cold feet and being susceptible to colds and chills during the winter and who, as a child, had wished the world would be covered with three inches of warm water—developed an "undersock" that would keep feet warm in the winter and cold during the summer. Burns, *John Logie Baird*, 18.

30. M. H. I. Baird, "Baird in America."

31. Ritchie, *Please Stand By*.

32. Burns, *John Logie Baird*, 112.

33. Burns, *John Logie Baird*, 112.

34. Burns, *John Logie Baird*, 116.

35. Troup, *Therapeutic Uses of Infra-red Rays*.

36. Woloshyn, *Soaking Up the Rays*.

37. "MM. Osty's Investigations."

38. "Provide Music by Radio," *Talking Machine World*, November 15, 1921, 26.

39. Liu, *The Laws of Cool*.

40. Horikoshi et al., *Microwave Chemical and Materials Processing*, 31.

41. "Short-Wave Tube," *Science News-Letter* 13, no. 361 (1928): 149.

42. "Ray Light Frozen," *Los Angeles Times*, December 5, 1930.

43. "Radio Produces Artificial Fever," 227.

44. "Radio Produces Artificial Fever," 227. Dr. W. T. Richards at Princeton and Alfred L. Loomis in Tuxedo Park, New York, undertook similar experiments.

45. Carpenter and Page, "Production of Fever in Man by Short Radio Waves," 450.

46. Kilduffe, "Synthetic Fever as a Treatment for Disease."

47. Schereschewsky, "Heating Effect."

48. J. H. Davis, "Radio Waves Kill Insect Pests," *Scientific American*, May 1933, 272.

49. "Wonders Done," *Los Angeles Times*, December 11, 1927.

50. "Wonders Done."

51. "Surgeons' Radio Knife Uses Human 'Antennae,'" *Washington Post*, March 10, 1931.

52. "The Fast-Moving Radio Field," *Radio Craft*, November 1936, 299.

53. "Heat Broadcasting," *New York Times*, January 24, 1926.

54. "Heat Broadcasting."

55. "Heat Broadcasting."

56. "Heat Broadcasting."

57. H. I. Phillips, "Heating by Radio," *Philadelphia Inquirer*, June 26, 1930.

58. Phillips, "Heating by Radio."

59. "Suggest Radio May Transmit Heat," *New York Times*, June 7, 1930.

60. "Suggest Radio May Transmit Heat."

61. "Radio Waves to Run Machinery," *Pittsburgh Gazette Times*, March 27, 1927.

62. "Cooking with Short Waves," *Short Wave Craft*, November 1933, 394.

63. "Cooking with Short Waves," 394.

64. "Radio Dealers Find Diathermy Profitable to Sell to Homes," *Radio Today* November 1938, 38.

65. "Human Body Shown to Radiate Energy," *New York Times*, March 29, 1934.

66. "Human Body Shown to Radiate Energy."

67. "Bell's Photophone," *Scientific American*, January 1, 1881, 4.

68. "Bell's Photophone," 4.

69. Burns, *Invasion of the Mind Snatchers*, 2.

70. Rao, *Optical Communication*, 1.

71. Li, preface, xi.

72. Goff and Hansen, *Fiber Optic Reference Guide*.

73. Goff and Hansen, *Fiber Optic Reference Guide*.

74. Goff and Hansen, *Fiber Optic Reference Guide*.

75. "Optical Fiber Loss and Attenuation," FOSCO: Fiber Optics for Sale Co., September 20, 2010, https://www.fiberoptics4sale.com/blogs/archive-posts/95048006-optical-fiber-loss-and-attenuation.

76. "Optical Fiber Loss and Attenuation."

77. The 850 nm, 1310 nm, and 1550 nm wavelengths are also chosen specifically to avoid "water bands" in glass fibers, which increase attenuation.

78. "Understanding Wavelengths in Fiber Optics," Fiber Optic Association, accessed September 3, 2019, https://www.thefoa.org/tech/wavelength.htm.

79. Goff and Hansen, *Fiber Optic Reference Guide*.

80. Goff and Hansen, *Fiber Optic Reference Guide*.

81. Goff and Hansen, *Fiber Optic Reference Guide*, 4.

82. Goff and Hansen, *Fiber Optic Reference Guide*, 4.

83. "Understanding Wavelengths in Fiber Optics."

84. Chun, *Control and Freedom*, 35.

85. Hecht, *City of Light*.

86. Brian W. Brush and Yong Ju Lee, "Filament Mind," Teton County Library, 2013.

87. *Merriam-Webster Online*, s.v. "light (*n.*)," accessed September 3, 2019, https://www.merriam-webster.com/dictionary/light.

88. Tyler et al., "Using Distributed Temperature Sensors."

89. Smolen and van der Spek, *Distributed Temperature Sensing*, 5.

90. Lindsay Tallon and Mike O'Kane, "Applying Distributed Temperature Sensing to the Heap Leach Industry," Heap Leach Conference, Vancouver, 2013, accessed November 15, 2015, https://www.okc-sk.com/au/wp-content/uploads/2015/04/Tallon-and-OKane-2013-Applying-distributed-temperature-sensing-to-the-heap-leach-industry.pdf.

91. Rogers, "Invited Paper Distributed Sensors."

92. Johnston and Kirkham, "An Electric Field Meter."

93. "Bioleaching Project in Chile," RocTest, March 2, 2017, https://roctest.com/en/case-study/bioleaching-project-in-chile/.

94. Kathy Kincade, "Environmental Monitoring: Distributed Temperature Sensing Aids Global-Warming Studies," *LaserFocusWorld*, December 1, 2006, https://www.laserfocusworld.com/test-measurement/research/article/16546493/environmental-monitoring-distributed-temperature-sensing-aids-globalwarming-studies.

95. Howe et al., "SMART Cables."

96. Ajo-Franklin et al., "Distributed Acoustic Sensing."

97. Jousset et al., "Dynamic Strain Determination."

98. Lindsey, Dawe, and Ajo-Franklin, "Illuminating Seafloor Faults."

99. Tawil-Souri, "Spectrum."

100. Witze, "Global 5G Wireless Deal."

Chapter 5: Infrared Camera

Portions of this chapter were previously published as "Thermal Vision," *Journal of Visual Culture* 18, no. 2 (2019): 1–23.

1. In a compelling analysis of the "aesthetics of cold," Luis Antunes reveals an array of ways that thermoceptive cues communicate in cinema, ranging from representations of changing states of matter to facial expressions to character movement. Antunes, "Thermoception in the Arctic Film."

2. Marks, *Skin of the Film*.

3. Lee, *Companion*, 51.

4. Mogensen and English, "Apparent Warmth of Colors"; Bennett and Rey, "What's So Hot about Red?"; Balcer et al., "Is Seeing Warm, Feeling Warm?"

5. Guéguen and Jacob, "Coffee Cup Color"; Ziat et al., "A Century Later." Aside from their demonstration of the synesthetic possibilities of thermoception, these investigations serve an ideological purpose as well. A better knowledge of how to visually activate thermal associations and sensations benefits companies interested in selling products not only via heating and air-conditioning but through the use of imagery.

6. Gabrys, "Sensors and Sensing Practices."

7. Parks, "Vertical Mediation," 143.

8. Agamben, *Homo Sacer*.

9. Parks, "Vertical Mediation," 145.

10. Langley, "The Bolometer and Radiant Energy."

11. Wood, "Photography by Invisible Rays," 329.

12. Wood, "Photography by Invisible Rays," 330.

13. Wood, "Photography by Invisible Rays," 336.

14. "Annual Exhibition of the Royal Photographic Society."

15. Rawling, *Infra-red Photography*, vii.

16. "Photos Taken in the Dark," *Popular Mechanics*, February 1932, 276.

17. Herbert Brennon, "Photographing the Invisible," *Camera Craft*, April 1933, 154.

18. Bawden, "Infra-red Photography," 168

19. Seligman, "Infra-red Photographs."

20. Jones, "Demonstration of Collateral Venous Circulation"; Ronchese, "Infrared Photography"; Epstein, "Infrared Photographic Demonstration."

21. Kantor, "Infrared Motion-Picture Technique." Some of these are described in Clarke's book, *Photography by Infrared*.

22. Cartwright, *Screening the Body*; van Dijck, *The Transparent Body*.

23. Irving and Hart, "Metabolism and Insulation of Seals"; Bailey and Davis, "Utilization of Body Fat."

24. J. R. Clark, "Thermal Pollution and Aquatic Life," *Scientific American*, March 1969, 18–27.

25. Lawson, Wlodek, and Webster, "Thermographic Assessment," 1129.

26. Lawson, Wlodek, and Webster, "Thermographic Assessment," 1129.

27. This is not always a clear-cut relationship. Different forms of matter also vary in emissivity, a material's ability to emit radiation. For this reason, to gauge

temperatures accurately, some encourage painting the surface of an object black—thus giving a "truer" temperature. Low-emissivity surfaces and substances, such as polished metals, are a kind of thermal mirror—they emit less radiation, and a thermal sensor that is calibrated to a high-emissivity substance, such as ice or brick, will record these as colder than they actually are. Rougher surfaces tend to reflect less infrared radiation than smooth ones. Black objects tend to emit more than white ones. On some thermal imaging cameras, the image must be calibrated so that the norm is either matte or glossy, thus offering always a skewed temperature of nonmatte or nonglossy objects. Thermal images are therefore never simply recordings of temperature but of interactions among objects, surfaces, and materialities in particular conditions, calibrated to particular norms.

28. At least this is the case in passive thermography. In active thermography, an energy source is used to "illuminate" the subject.

29. Jacobs, *Panic at the Pump*.

30. Marshall, *Infrared Thermography of Buildings*, iv.

31. H. Shuldiner, "Infrared Scanners Can Help Cut Your Home Energy Bills," *Popular Science*, September 1975, 86, 132.

32. Other models included the Barnes Model T-101, Barnes IAX-8, Bendix M2S, ERIM M7, Inframetrics Model 510, and Texas Instruments B-310, AN/AAS-18. Marshall, *Infrared Thermography of Buildings*.

33. Shuldiner, "Infrared Scanners Can Help."

34. H. Shuldiner, "Heat-Leak Locator: Thermographic Scanner Charts Insulation Gaps," *Popular Science*, January 1978, 83.

35. US Department of Agriculture, Foreign Agricultural Service, "GLAM—Global Agricultural Monitoring," accessed April 4, 2021, https://ipad.fas.usda.gov/glam .htm.

36. Chris Anderson, "Ten Lessons for Farm Drones," *Robohub*, October 22, 2013, http://robohub.org/ten-lessons-for-farm-drones/.

37. Peter Murray, "Drones Close In on Farms, the Next Step in Precision Agriculture," *Precision Farming Dealer*, May 28, 2013, http://www.precisionfarmingdealer .com/articles/405-drones-close-in-on-farms-the-next-step-in-precision -agriculture#sthash.GfxJ2Uxj.dpuf.

38. David Herring, "Precision Farming," NASA Earth Observatory, accessed April 4, 2021, https://earthobservatory.nasa.gov/features/PrecisionFarming /precision_farming2.php.

39. Anderson, "Ten Lessons for Farm Drones."

40. Miles, "The Combine Will Tell the Truth," 8.

41. Amstrup et al., "Detecting Denning Polar Bears"; Bernatas and Nelson, "Sightability Model"; Campbell and Donlan, "Feral Goat Eradications on Islands."

42. "The Tasmanian Tiger Down Under," *Extinct or Alive*, season 1, episode 6, Animal Planet, July 15, 2018.

43. "Wildlife Crime Technology Project," World Wildlife Fund, accessed December 22, 2018, https://www.worldwildlife.org/projects/wildlife-crime-technology -project.

44. "Wildlife Crime Technology Project."

45. Travis Merrill, "FLIR Works with World Wildlife Fund to Fight Poaching in Africa," FLIR, November 20, 2016, https://www.flir.com/news-center/corporate -news/flir-works-with-world-wildlife-fund-to-fight-poaching-in-africa.

46. Carter Roberts, "Applying Technology to Our Conservation Mission," *Medium*, June 25, 2016, https://medium.com/@Carter_Roberts/applying-technology -to-our-conservation-mission-8101a4a0d9ec; Clair MacDougall, "In Africa, Technology Is the Final Weapon in the Deadly Poaching War," *Wired*, May 2, 2018, https:// www.wired.co.uk/article/anti-poaching-technology-conservation-maasai-mara; James Morgan, "Using Technology to Combat Wildlife Crime," National Geographic Society Newsroom, November 21, 2016, https://blog.nationalgeographic.org/2016 /11/21/using-technology-to-combat-wildlife-crime.

47. For a history of camera traps, especially those that use infrared, see O'Connell, Nichols, and Karanth, *Camera Traps in Animal Ecology*.

48. Lunstrum, "Green Militarization."

49. *Proceedings: Protecting Threatened Wildlife in Africa with Technology and Training: A Forum Hosted by the Richardson Center for Global Engagement*, Richardson Center for Global Engagement, World Wildlife Fund, and African Parks, October 31, 2013, http://www.richardsondiplomacy.org/wp-content/themes/digisans/images /PROCEEDINGS_WWF_RICHARDSON_AP_FORUM_FINAL.pdf.

50. "TeleEye Unveils Thermal Camera to Combat Rhino Poaching," *Security Solutions Magazine*, October 2015, http://www.securitysa.com/53073n.

51. Secu-Systems, "Systems containers complete integration with FLIR and access containers . . . self sustaining green energy," Facebook, March 9, 2016, https:// www.facebook.com/SecuSystemsAfrica/posts/753865414749438; "Stallion Secures Renewable Energy Plant in Contract Valued at R11 Million," *Security Solutions Magazine*, April 2014, http://www.securitysa.com/6433r.

52. Jack Shainman Gallery, "Richard Mosse, *Heat Maps*, February 2–March 11, 2017," press release, accessed January 30, 2021, https://jackshainman.com /exhibitions/heat_maps.

53. Jack Shainman Gallery, "Richard Mosse."

54. "Horizon," Leonardo, accessed December 11, 2018, https://www .leonardocompany.com/en/-/horizon-1.

55. Marks, *Skin of the Film*.

56. Mosse, "Transmigration of the Souls," 1.

57. Mosse, "Transmigration of the Souls," 2.

58. Bryant, "Letter from a Sensor Operator," 320.

59. Mosse, "Transmigration of the Souls," 2.

60. Martin, "As 'Index and Metaphor,'" 9.

61. Parks, "Vertical Mediation," 145.

62. Parks, "Vertical Mediation," 145.

63. Wood, "Photography by Invisible Rays," 336.

64. Roaf, Brotas, and Nicol, "Counting the Costs of Comfort," 272.

65. Albers, Maier, and Marggraf-Micheel, "In Search of Evidence," 492.

Portions of this chapter were previously published as "Thermocultures of Geological Media," *Cultural Politics* 12, no. 3 (2016): 293–309.

1. McPherson, "U.S. Operating Systems."

2. Noon, "Negotiating a Quantum Computation Network," 125, 136.

3. "Quantum Computing at IBM," IBM Q, accessed February 1, 2019, https://www.research.ibm.com/ibm-q/learn/what-is-ibm-q.

4. Noon, "Negotiating a Quantum Computation Network," 138.

5. Noon, "Negotiating a Quantum Computation Network," 143.

6. "Quantum Computing at IBM."

7. "Quantum Computing at IBM."

8. Brunton, "Heat Exchanges," 159.

9. Velkova, "Data That Warms."

10. Rubio, *Still Life*. For an analysis of the ways that memory institutions are dependent on thermal media, see Bhowmik, "Thermocultures of Memory."

11. Radin and Kowal, "The Politics of Low Temperature," 7.

12. Friedrich Nietzsche used the language of cooling to propose a mode of subduing human desires, particularly the will to believe in some transcendental agency: "Do not deride and befoul that which you want to do away with for good, but respectfully *lay it on ice*." Freezing is a philosophical tactic that halts investments in the metaphysical. Nietzsche, "Wanderer and His Shadow," 361.

13. Radin and Emma Kowal, "The Politics of Low Temperature," 7.

14. For a critical and compelling discussion of the frozen archive, see Hogan and Roberts, "Archives Melting (and Meltdowns)."

15. For GilFillan, the "leadership in world civilization is inseparably linked with climate." The "positive value" of warmth was necessary for the development of productive agriculture, freed workers up for other activities, and made civilization possible. GilFillan, "Coldward Course of Progress," 393.

16. Velkova, "Data Centers as Impermanent Infrastructures."

17. "Practical Talks on Presswork," *Inland Printer* 8, no. 4 (1891): 293.

18. "Practical Talks on Presswork," 293.

19. Allan Turck, "What the Color Printer Is up Against," *Inland Printer* 53, no. 2 (April 1914): 222.

20. "The Pressfeeder," *Inland Printer* 53 (1914): 67.

21. "Even Temperature in the Pressroom," *Inland Printer* 29, no. 3 (June 1902): 394.

22. "Even Temperature," 394.

23. Turck, "What the Color Printer," 221.

24. Colebrook, "Our London Letter," 211.

25. "The Invention That Changed the World," Willis Carrier, accessed February 19, 2019, http://www.willliscarrier.com/1876–1902.php.

26. Paper production continues to be tightly connected to thermal technologies even today, including not only air conditioners but also wet-end steamboxes, which reduce moisture variance.

27. Ingels, *Willis H. Carrier*, 34.

28. Baker, "Air Cleaning and Conditioning," 5.

29. Simonds and Polderman, "Air Conditioning in Film Laboratories," 607.

30. Baker, "Air Cleaning and Conditioning," 5.

31. Simonds and Polderman, "Air Conditioning in Film Laboratories," 607.

32. Simonds and Polderman, "Air Conditioning in Film Laboratories," 611.

33. Simonds and Polderman, "Air Conditioning in Film Laboratories," 607.

34. "Universal Studio in East Vies with City in West," *Motion Picture News*, June 26, 1915, 57.

35. Harry A. Mount, "Making Weather to Order," *Scientific American*, March 5, 1921, 188, 198–99.

36. Simonds and Polderman, "Air Conditioning in Film Laboratories," 607.

37. "The Newest of Motion Picture Plants," *Motion Picture News*, December 27, 1913, 30.

38. "Newest of Motion Picture Plants," 30.

39. Simonds and Polderman, "Air Conditioning in Film Laboratories," 612.

40. Kiesling, *Talking Pictures*, 5.

41. Simonds and Polderman, "Air Conditioning in Film Laboratories," 607.

42. "Fuel Denied," *Motion Picture News*, January 19, 1918, 389.

43. "New York or Los Angeles?," *Camera*, May 4, 1919, 4.

44. Simonds and Polderman, "Air Conditioning in Film Laboratories," 605–6.

45. Simonds and Polderman, "Air Conditioning in Film Laboratories," 607–8.

46. Simonds and Polderman, "Air Conditioning in Film Laboratories," 605.

47. "5-Cent Show Easily Started," *Moving Picture World*, August 17, 1907, 376.

48. Baker, "Air Cleaning and Conditioning," 4.

49. Carlile, *Production and Direction*, 349.

50. Barnouw, *A Tower in Babel*, 194.

51. "Studio Tours Begin in Hollywood Radio City," NBC *Transmitter*, January 1939, 1.

52. National Broadcasting Company, "The Weather Eats out of His Hand," *Radio Daily*, August 11, 1944, 3.

53. National Broadcasting Company, "The Weather Eats out of His Hand," 63.

54. "Developing Our Export Trade," *Talking Machine World*, February 15, 1908, 18.

55. "Developing Our Export Trade," 18.

56. "From Our European Headquarters," *Talking Machine World*, February 1907, 25.

57. "From Our European Headquarters," 25.

58. "Larger Cylinder Records," *Talking Machine World, July 1906*, 6.

59. "Latest Patents Relating to Talking Machines and Records," *Talking Machine World*, May 1914, 60.

60. "Latest Patents Relating to Talking Machines and Records," 65.

61. "U.S. Everlasting Non-Breakable Records," *Talking Machine World*, June 1911, 25.

62. Andrew H. Dodin, "Watch for Chilled Main Springs," *Talking Machine World*, October 15, 1923, 184.

63. "Use Vulcan Mainsprings," *Talking Machine World*, July 15, 1925, 173.

64. "Steurer Reproducer Company," *Talking Machine World*, December 15, 1920, 188.

65. "Reports Good Trade Prospects," *Talking Machine World*, September 1916, 94.

66. "Trade News from All Points of the Compass," *Talking Machine World*, June 1905, 24.

67. "Warping of Records," *Talking Machine World*, December 1905, 11.

68. Stefan Michalski, "Agent of Deterioration: Incorrect Temperature," Canadian Conservation Institute, accessed April 4, 2021, https://www.canada.ca/en /conservation-institute/services/agents-deterioration/temperature.html.

69. Rubio, "Preserving the Unpreservable," 620.

70. Robert Gannon, "Life in a Germfree World," *Popular Science*, August 1962, 92.

71. Tatiana Schlossberg, "An Ice Scientist's Worst Nightmare," *New York Times,* April 11, 2017.

72. Radin and Kowal, "The Politics of Low Temperature,"6.

73. Terence McGinley, "Watch the Cyan! How The New York Times Gets Inked," *New York Times*, May 23, 2017.

74. "Paper Mill," Thermoelectric Cooling America Corporation, accessed February 18, 2019, https://www.thermoelectric.com/paper-mill/.

75. Mark Gregory, "Inside Facebook's Green and Clean Arctic Data Centre," BBC *News*, June 14, 2013; Ned Potter, "Facebook Plans Server Farm in Sweden; Cold Is Great for Servers," ABC *News*, October 27, 2011; Eoghan Macguire, "Can Scandinavia Cool the Internet's Appetite for Power?," CNN, November 16, 2014.

76. Michael Shalyt, "Data: An Achilles' Heel in the Grid?," Power Engineering International, May 22, 2017, https://www.powerengineeringint.com/digitalization /data-an-achilles-heel-in-the-grid/.

77. Cara Garretson, "Stupid Data Center Tricks," *InfoWorld*, August 12, 2010, https://www.infoworld.com/article/2625613/servers/stupid-data-center-tricks.html.

78. Zoë Schlanger, "What Happened to the 'Fail-Safe' Svalbard Seed Vault Designed to Save Us from Crop Failure" *Quartz*, May 20, 2017, https://qz.com/987894 /the-fail-safe-svalbard-seed-vault-designed-to-save-us-from-crop-failure-just -flooded-thanks-to-climate-change.

79. Rich Miller, "ASHRAE: Warmer Data Centers Good for Some, Not All," *Data Center Knowledge*, October 5, 2012, http://www.datacenterknowledge.com/archives /2012/10/05/beaty-ashrae-temperature.

80. Velkova, "Data That Warms," 1.

81. Sanjay Bhatt, "Amazon, Data Center Turn Hot Idea into Cool Technology," *Seattle Times*, November 12, 2015.

82. "The Super-Efficient Heat Source Hidden below Amazon's Seattle Headquarters," *Day One: The Amazon Blog*, November 16, 2017, https://blog.aboutamazon .com/sustainability/the-super-efficient-heat-source-hidden-below-amazons-new -headquarters.

83. Alger, *Grow a Greener Data Center*.

84. Rich Miller, "Study: Server Failures Don't Rise along with the Heat," *Data Center Knowledge*, May 29, 2012, http://www.datacenterknowledge.com/archives /2012/05/29/study-server-failures-dont-rise-along-with-the-heat.

85. Andrew Donoghue, "Data Center Cooling: It's Time to Go with the (Air) Flow," *Verne Global*, April 4, 2021, https://verneglobal.com/news/blog/data-center -cooling-its-time-to-go-with-the-air-flow.

86. Donoghue, "Data Center Cooling."

87. Wiener, *Cybernetics*, 96.

88. One book on social cognitive psychology puts forth, for example, that "principles of cybernetics" are best described through "the example of a simple and familiar self-regulating system: the thermostat." Barone, Maddux, and Snyder, *Social Cognitive Psychology*, 279.

89. Wishart, "Honeywell," 126.

90. "Is Cooling the Top Efficiency Metric for Tropical Data Centers?," Pergravis, April 26, 2016, http://pergravis.com/industry-insights/blog/is-cooling-the-top -efficiency-metric-for-tropical-data-centers/.

91. "Is Cooling the Top Efficiency Metric for Tropical Data Centers?"

92. Peter Judge, "The Hottest and Coolest Data Center Locations," *Data Center Dynamics*, August 29, 2018, https://www.datacenterdynamics.com/analysis/hottest -and-coolest-data-center-locations/.

93. Judge, "Hottest and Coolest."

94. Jiao et al., "Thermal Analysis," 224.

95. Jiao et al., "Thermal Analysis"; Alyssa Danigelis, "Efficient Tropical Data Center High-Rise Eyed for Singapore," Environment and Energy Leader, August 25, 2017, https://www.environmentalleader.com/2017/08/efficient-tropical-data-center -high-rise-eyed-singapore/.

96. Paulo Cesar de Resende Pereira, "Tropical Sites Need Solar Power, Not Free Cooling," *Data Center Dynamics*, March 24, 2016, https://www.datacenterdynamics .com/analysis/tropical-sites-need-solar-power-not-free-cooling/.

97. Danigelis, "Efficient Tropical Data Center."

98. Judge, "Hottest and Coolest."

99. Fred Pearce, "The Data on Green Data Centers Is Still Pretty Cloudy," *GreenBiz*, April 24, 2018, https://www.greenbiz.com/article/data-green-data-centers-still -pretty-cloudy.

100. McLuhan, *Understanding Media*.

101. Wishart, "Honeywell," 125–26.

102. Wishart, "Honeywell," 126.

Conclusion

1. Janet Walker, "Afterword," 307.

2. Howe and Boyer, "Redistributions."

3. Anderson, *After the Ice*; Pollack, *A World without Ice*.

4. Marx and Engels, *Manifesto of the Communist Party*, 16.

5. Take, for example, the early twentieth-century manufacture of records. First, raw material was moved through steam-heated rollers, "which raise[d] the composition to the exact temperature at which it develop[ed] the required degree of softness" and became "an absolutely smooth, plastic mass." This "black dough" was rolled out into strips and conveyed by a canvas belt into a cooling chamber. After this it was sent to the pressing floors, where the squares of record material were heated on a steam table to "just the right temperature" and imprinted with a given label. "C. H. Wicks, Superintendent Record Pressing Plant," *Talking Machine World*, July 15, 1916, 43.

6. Hogan and Roberts, "Archives Melting (and Meltdowns)."

7. Whyte, "Indigenous Science (Fiction)."

8. Yusoff, "Epochal Aesthetics."

9. See Balsamo, *Technologies of the Gendered Body*.

10. Peters, *The Marvelous Clouds*.

Abe, Yohei, Yosuke Fujiwara, Hiroki Takahashi, Yoshihiro Matsumura, Tomonobu Sawada, Shuying Jiang, Ryo Nakaki, et al. "Histone Demethylase JMJD1A Coordinates Acute and Chronic Adaptation to Cold Stress via Thermogenic Phospho-switch." *Nature Communications* 9 (2018): 1–16.

Ackermann, Marsha E. *Cool Comfort: America's Romance with Air-Conditioning.* Washington, DC: Smithsonian Books, 2010.

Adams, Sean Patrick. *Home Fires: How Americans Kept Warm in the Nineteenth Century.* Baltimore: Johns Hopkins University Press, 2014.

Agamben, Giorgio. *Homo Sacer: Sovereign Power and Bare Life.* Translated by Daniel Heller-Roazen. Stanford, CA: Stanford University Press, 1998.

Agbemabiese, Lawrence, Kofi Berko Jr., and Peter du Pont. "Air Conditioning in the Tropics: Cool Comfort or Cultural Conditioning?" In *Proceedings of the 1996 Summer Study on Energy Efficiency in Buildings*, 8.1–8.9. Washington, DC: American Council for an Energy Efficient Economy, 1996.

Ajo-Franklin, Jonathan B., Shan Dou, Nathaniel J. Lindsey, Inder Monga, Chris Tracy, Michelle Robertson, Veronica Rodriguez Tribaldos, et al. "Distributed Acoustic Sensing Using Dark Fiber for Near-Surface Characterization and Broadband Seismic Event Detection." *Scientific Reports* 9, no. 1328 (2019): 1–14. https://doi.org/10.1038/s41598-018-36675-8.

Alaimo, Stacy. *Bodily Natures: Science, Environment, and the Material Self.* Bloomington: Indiana University Press, 2010.

Alaimo, Stacy. *Exposed: Environmental Politics and Pleasures in Posthuman Times.* Minneapolis: University of Minnesota Press, 2016.

Albers, F., J. Maier, and C. Marggraf-Micheel. "In Search of Evidence for the Hue-Heat Hypothesis in the Aircraft Cabin." *Lighting Research and Technology* 47, no. 4, (2015): 483–94.

Aldrich, Mark. "An Energy Transition before the Age of Oil: The Decline of Anthracite, 1900–1930." *Pennsylvania History: A Journal of Mid-Atlantic Studies* 85, no. 1 (2018): 1–31.

Alger, Douglas. *Grow a Greener Data Center.* Indianapolis: CISCO, 2010.

Allen, John R. "Modus Operandi of the Synthetic Air Chart." *Transactions: American Society of Heating and Ventilating Engineers* 26, no. 6 (1922): 597–603.

Allen-Collinson, Jacquelyn, and Helen Owton. "Intense Embodiment: Senses of Heat in Women's Running and Boxing." *Body and Society* 21, no. 2 (2015): 245–68.

Allen-Collinson, Jacquelyn, Anu Vaittinen, George Jennings, and Helen Owton. "Exploring Lived Heat, 'Temperature Work,' and Embodiment: Novel Auto/Ethnographic Insights from Physical Cultures." *Journal of Contemporary Ethnography* 47, no. 3 (2018): 283–305.

Alter, Nora M., Lutz Koepnick, and Richard Langston. "Landscapes of Ice, Wind, and Snow: Alexander Kluge's Aesthetic of Coldness." *Grey Room*, no. 53 (2013): 60–87.

American Society of Heating and Ventilating Engineers. *American Society of Heating and Ventilating Engineers Guide*. New York: American Society of Heating and Ventilating Engineers, 1924–1925.

American Society of Heating and Ventilating Engineers. Standard 55-2013. *Thermal Environmental Conditions for Human Occupancy*. Atlanta: ASHRAE, 2013.

American Society of Heating and Ventilating Engineers. Standard 55-2017. *Thermal Environmental Conditions for Human Occupancy*. Atlanta: ASHRAE, 2017.

Amstrup, Steven C., Geoff York, Trent L. McDonald, Ryan Nielson, and Kristin Simac. "Detecting Denning Polar Bears with Forward-Looking Infrared (FLIR) Imagery." *BioScience* 54, no. 4 (2004): 337–44.

Anderson, Alun. *After the Ice: Life, Death, and Geopolitics in the New Arctic*. Washington, DC: Smithsonian Books, 2009.

Antunes, Luis. "Thermoception in the Arctic Film: Knut Erik Jensen's 'Aesthetics of Cold.'" *Cine-Files* 10 (2016). http://www.thecine-files.com/antunes2016/.

Aristotle. *On the Parts of Animals, Book II*. Translated by William Ogle. London: Kegan Paul, French and Co., 1882.

Atanasoski, Neda, and Kalindi Vora. *Surrogate Humanity: Race, Robots, and the Politics of Technological Futures*. Durham, NC: Duke University Press, 2019.

Bachelard, Gaston. *The Psychoanalysis of Fire*. Translated by Alan C. M. Ross. Boston: Beacon, 1964.

Bailey, E. D., and D. E. Davis. "The Utilization of Body Fat during Hibernation in Woodchucks." *Canadian Journal of Zoology* 43, no. 5 (1965): 701–7.

Baker, T. Thorne. "Air Cleaning and Conditioning." *Proceedings of the British Kinematograph Society*, no. 15 (1933): 3–12.

Balsamo, Anne. *Technologies of the Gendered Body: Reading Cyborg Women*. Durham, NC: Duke University Press, 1995.

Barad, Karen. *Meeting the Universe Halfway: Quantum Physics and the Entanglement of Matter and Meaning*. Durham, NC: Duke University Press, 2007.

Barber, Daniel A. *Modern Architecture and Climate: Design before Air Conditioning*. Princeton, NJ: Princeton University Press, 2020.

Barnouw, Erik. *A Tower in Babel*. Vol. 1 of *A History of Broadcasting in the United States to 1933*. New York: Oxford University Press, 1966.

Barone, David F., James E. Maddux, and C. R. Snyder. *Social Cognitive Psychology: History and Current Domains*. New York: Plenum, 1997.

Basile, Salvatore. *Cool: How Air Conditioning Changed Everything*. New York: Fordham University Press, 2014.

Bate, David G., and Andrew L. Morrison. "Some Aspects of the British Geological Survey's Contribution to the War Effort at the Western Front, 1914–1918." *Proceedings of the Geologists' Association* 129, no. 1 (2018): 3–11.

Bawden, F. C. "Infra-Red Photography and Plant Virus Diseases." *Nature* 132, no. 3326 (1933): 168.

Bedini, Silvio A. "The Role of Automata in the History of Technology." *Technology and Culture* 5, no. 1 (1964): 24–42.

Bell, Alexander Graham. "Selenium and the Photophone." *Nature* 22 (1880): 500–503.

Bell, Alexander Graham. "Upon the Production of Sound by Radiant Energy." *Proceedings of American Association for the Advancement of Science* (August 27, 1880): 1–45.

Bennett, Corwin A., and Paule Rey. "What's So Hot about Red?" *Human Factors: The Journal of the Human Factors and Ergonomics Society* 14, no. 2 (1972): 149–54.

Beregow, Elena, ed. "Thermal Objects" (special issue). *Culture Machine* 17 (2018).

Beregow, Elena. "Thermal Objects: Theorizing Temperatures and the Social." *Culture Machine* 17 (2018): 1–18.

Beretta, Davide, Neophytos Neophytou, James M. Hodges, Mercouri G. Kanatzidis, Dario Narducci, Marisol Martin-Gonzalez, Matt Beekman, et al. "Thermoelectrics: From History, a Window to the Future." *Materials Science and Engineering: R: Reports* 138 (2019): 210–55.

Berlant, Lauren, and Lee Edelman. *Sex, or the Unbearable*. Durham, NC: Duke University Press, 2014.

Bernatas, Susan, and Lou Nelson. "Sightability Model for California Bighorn Sheep in Canyonlands Using Forward-Looking Infrared (FLIR)." *Wildlife Society Bulletin* 32, no. 3 (2004): 638–47.

Bhowmik, Samir. "Thermocultures of Memory." *Culture Machine* 17 (2018): 1–20.

Blum, Hester. *The News at the Ends of the Earth: The Print Culture of Polar Exploration*. Durham, NC: Duke University Press, 2019.

Bollinger, Martin J. *Stalin's Slave Ships: Kolyma, the Gulag Fleet, and the Role of the West*. Westport, CT: Praeger, 2008.

Boyer, Dominic. *Energopolitics: Wind and Power in the Anthropocene*. Durham, NC: Duke University Press, 2019.

Bradshaw, Vaughn. *The Building Environment: Active and Passive Control Systems*. Hoboken, NJ: John Wiley and Sons, 2006.

Bruce, Robert V. *Alexander Graham Bell and the Conquest of Solitude*. Ithaca, NY: Cornell University Press, 1973.

Brunton, Finn. "Heat Exchanges." In *The MoneyLab Reader: An Intervention in Digital Economy*, edited by Geert Lovink, Nathaniel Tkacz, and Patricia de Vries, 158–73. Amsterdam: Institute of Network Cultures, 2015.

Bryant, Brandon. "Letter from a Sensor Operator." In *Life in the Age of Drone Warfare*, edited by Lisa Parks and Caren Kaplan, 315–23. Durham, NC: Duke University Press, 2017.

Buensod, A. C. "Theater Air Conditioning." In *The 1931 Film Daily Year Book of Motion Pictures*, edited by Jack Alicoate, 539–40. New York: Film Daily, 1931.

Burns, Eric. *Invasion of the Mind Snatchers: Television's Conquest of America in the Fifties*. Philadelphia: Temple University Press, 2010.

Burns, Russell W. *John Logie Baird: Television Pioneer*. London: Institution of Engineering and Technology, 2000.

Cabanac, Michel. "Sensory Pleasure." *Quarterly Review of Biology* 54, no. 1 (1979): 1–29.

Campbell, Karl, and C. Josh Donlan. "Feral Goat Eradications on Islands." *Conservation Biology* 19, no. 5 (2005): 1362–74.

Candido, Christhina, and Richard de Dear. "From Thermal Boredom to Thermal Pleasure: A Brief Literature Review." *Ambiente Construído* 12, no. 1 (2012): 81–90.

Cantor, Guillermo. *Hieleras (Iceboxes) in the Rio Grande Valley Sector: Lengthy Detention, Deplorable Conditions, and Abuse in CBP Holding Cells*. American Immigration Council, December 17, 2015. https://www.americanimmigrationcouncil.org /research/hieleras-iceboxes-rio-grande-valley-sector.

Carlile, John Snyder. *Production and Direction of Radio Programs*. New York: Prentice-Hall, 1939.

Carpenter, Charles M., and Albert B. Page. "Production of Fever in Man by Short Radio Waves." *Science*, May 2, 1930, 450–52.

Cartwright, Lisa. *Screening the Body: Tracing Medicine's Visual Culture*. Minneapolis: University of Minnesota Press, 1995.

Case, T. W. "Infra Red Telegraphy and Telephony." *Journal of the Optical Society of America* 6 (1922): 398–406.

Central Intelligence Agency. *Counterterrorism Detention and Interrogation Activities (September 2001–October 2003)*. Report no. 2003–7123-IG, May 7, 2004.

Chang, Hasok. *Inventing Temperature: Measurement and Scientific Progress*. New York: Oxford University Press, 2007.

Chang, Tom Y., and Agne Kajackaite. "Battle for the Thermostat: Gender and the Effect of Temperature on Cognitive Performance." *PLOS One* 14, no. 5 (2019). https://doi.org/10.1371/journal.pone.0216362.

Chol-hwan, Kang, and Pierre Rigoulot. *The Aquariums of Pyongyang: Ten Years in the North Korean Gulag*. New York: Perseus, 2000.

Chun, Wendy Hui Kyong. *Control and Freedom: Power and Paranoia in the Age of Fiber Optics*. Cambridge, MA: MIT Press, 2006.

Chun, Wendy Hui Kyong. "On Patterns and Proxies, or the Perils of Reconstructing the Unknown." *e-flux Architecture*, September 25, 2018. https://www.e-flux.com /architecture/accumulation/212275/on-patterns-and-proxies/.

Clark, Nigel. "Infernal Machinery: Thermopolitics of the Explosion." *Culture Machine* 17 (2018): 1–19.

Classen, Constance. *Worlds of Sense: Exploring the Senses in History and across Cultures*. London: Routledge, 1993.

Colebrook, Frank. "Our London Letter." *American Printer* 60, no. 2 (1915): 211.

Comstock, W. Stephen. "A Long, Strong Pull Together." In *Proclaiming the Truth: An Illustrated History of ASHRAE*, 1–16. Atlanta: ASHRAE, 2020.

Cooper, Gail. *Air-Conditioning America: Engineers and the Controlled Environment, 1900–1960*. Baltimore: Johns Hopkins University Press, 2002.

Cowen, Ruth Schwartz. *More Work for Mother: The Ironies of Household Technology from the Open Hearth to the Microwave*. New York: Basic Books, 1983.

Daggett, Cara New. *The Birth of Energy: Fossil Fuels, Thermodynamics, and the Politics of Work*. Durham, NC: Duke University Press, 2019.

Davis, Angela. "Reflections on the Black Woman's Role in the Community of Slaves." *Massachusetts Review* 13, nos. 1–2 (1972): 81–100.

Davis, Heather, and Zoe Todd. "On the Importance of a Date, or Decolonizing the Anthropocene." *ACME: An International Journal for Critical Geographies* 16, no. 4 (2017): 761–80.

Day, Deanna. "98.6: Fevers, Fertility, and the Patient Labor of American Medicine." PhD diss., University of Pennsylvania, 2014.

DeBow, J. D. B. "Ice—How Much of It Is Used, and Where It Comes From." *DeBow's Review* 19, no. 6 (December 1855): 709–12.

de Dear, Richard J., and Gail Schiller Brager. "Developing an Adaptive Model of Thermal Comfort and Preference." *ASHRAE Transactions* 104, part 1 (1998): 1–18.

De León, Jason. *The Land of Open Graves: Living and Dying on the Migrant Trail*. Berkeley: University of California Press, 2015.

Deleuze, Gilles, and Félix Guattari. *A Thousand Plateaus: Capitalism and Schizophrenia*. Translated by Brian Massumi. London: Continuum, 2004.

Diesen, Jan Anders. "A Century of Polar Expedition Films: From Roald Amundsen to Børge Ousland." In *Small Country, Long Journeys: Norwegian Expedition Films*, edited by Eirik Frisvold Hanssen and Maria Fosheim Lund, 83–115. Oslo: Nasjonalbiblioteket, 2017.

Dillon, Stephen. "Possessed by Death: The Neoliberal-Carceral State, Black Feminism, and the Afterlife of Slavery." *Radical History Review*, no. 112 (2012): 113–25.

Douglas, Mary. "Environments at Risk." In *Implicit Meanings: Selected Essays in Anthropology*, 230–48. New York: Routledge, 1975.

Epstein, Bernard S. "Infrared Photographic Demonstration of the Superficial Venous Pattern in Congenital Heart Disease with Cyanosis." *American Heart Journal* 18, no. 3 (1939): 282–89.

Equiano, Olaudah. "The Interesting Narrative of the Life of Olaudah Equiano, or Gustavus Vassa, the African, Written by Himself." In *Slave Narratives*, edited by William L. Andrews, 35–242. 1789. Reprint, New York: Library Classics of the United States, 2000.

Ernst, Wolfgang. "Time, Temperature and Its Informational Turn." *Culture Machine* 17 (2018): 1–13.

Esposito, Barbara, and Lee Wood. *Prison Slavery*. Edited by Kathryn Bardsley. Silver Spring, MD: Joel Lithographic, 1982.

Fanger, Povl Ole. *Thermal Comfort: Analysis and Applications in Environmental Engineering*. Copenhagen: Danish Technical Press, 1970.

Fennell, Catherine. *Last Project Standing: Civics and Sympathy in Post-Welfare Chicago*. Minneapolis: University of Minnesota Press, 2015.

Ferguson, Kevin L. "Panting in the Dark: The Ambivalence of Air in Cinema." *Camera Obscura: Feminism, Culture, and Media Studies* 77, no. 2 (2011): 32–63.

Fernández-Galiano, Luis. *Fire and Memory: On Architecture and Energy*. Translated by Gina Cariño. Cambridge, MA: MIT Press, 2000.

Foote, Christopher L., Warren C. Whatley, and Gavin Wright. "Arbitraging a Discriminatory Labor Market: Black Workers at the Ford Motor Company, 1918–1947." *Journal of Labor Economics* 21, no. 3 (2003): 493–532.

Fretwell, Erica. "Introduction: Common Senses and Critical Sensibilities." *Resilience: A Journal of the Environmental Humanities* 5, no. 3 (2018): 1–9.

Furuhata, Yuriko. *Climatic Media: Transpacific Experiments in Atmospheric Control*. Durham, NC: Duke University Press, 2022.

Gabrys, Jennifer. "Sensors and Sensing Practices: Reworking Experience across Entities, Environments, and Technologies." *Science, Technology, and Human Values* 44, no. 5 (2019): 723–36.

Gaines, Jane. "From Elephants to Lux Soap: The Programming and 'Flow' of Early Motion Picture Exploitation." *Velvet Light Trap* 25 (1990): 29–43.

GilFillan, S. C. "The Coldward Course of Progress." *Political Science Quarterly* 35, no. 3 (1920): 393–410.

Gilmore, Ruth Wilson. *Golden Gulag: Prisons, Surplus, Crisis, and Opposition in Globalizing California*. Berkeley: University of California Press, 2007.

Giovacchini, Saverio. *Hollywood Modernism: Film and Politics in the Age of the New Deal*. Philadelphia: Temple University Press, 2001.

Gitelman, Lisa. *Always Already New: Media, History, and the Data of Culture*. Cambridge, MA: MIT Press, 2006.

Goff, David R., and Kimberly S. Hansen. *Fiber Optic Reference Guide: A Practical Guide to Communications Technology*. Burlington, MA: Focal Press, 2013.

Goldsmith, Oliver. *The Vicar of Wakefield*. New York: Hurst, 1766.

Guéguen, Nicolas, and Cèline Jacob. "Coffee Cup Color and Evaluation of a Beverage's 'Warmth Quality.'" *Color Research and Application* 39, no. 1 (2014): 79–81.

Gumbs, Alexis Pauline. *M Archive: After the End of the World*. Durham, NC: Duke University Press, 2018.

Günel, Gökçe. *Spaceship in the Desert: Energy, Climate Change, and Urban Design in Abu Dhabi*. Durham, NC: Duke University Press, 2019.

Gunning, Tom. "Before Documentary: Early Nonfiction Films and the 'View' Aesthetic." In *Uncharted Territory: Essays on Early Nonfiction Film*, edited by Daan Hertogs and Nico De Klerk, 9–24. Amsterdam: Nederlands Filmmuseum, 1997.

Hansen, Mark B. N. *Feed-Forward: On the Future of Twenty-First-Century Media*. Chicago: University of Chicago Press, 2014.

Hartman, Saidiya. *Lose Your Mother: A Journey along the Atlantic Slave Route*. New York: Farrar, Straus and Giroux, 2007.

Harvey, Tom, Michael Patterson, and John Bean, eds. "Updated Air-Side Free Cooling Maps: The Impact of ASHRAE 2011 Allowable Ranges." The Green Grid, white

paper #46. Accessed February 19, 2019. https://datacenters.lbl.gov/sites/all/files
/WP46UpdatedAirsideFreeCoolingMapsTheImpactofASHRAE2011AllowableRan
ges.pdf.

Hecht, Jeff. *City of Light: The Story of Fiber Optics*. New York: Oxford University
Press, 1999.

Heijs, Wim, and Peter Stringer. "Research on Residential Thermal Comfort: Some
Contributions from Environmental Psychology." *Journal of Environmental Psychology* 8, no. 3 (1988): 237.

Hensley, Daniel W., Andrew E. Mark, Jayvee R. Abella, George M. Netscher,
Eugene H. Wissler, and Kenneth R. Diller. "50 Years of Computer Simulation of
the Human Thermoregulatory System." *Journal of Biomechanical Engineering* 135,
no. 2 (2013): 1–9.

Herold, Marc W. "Ice in the Tropics: The Export of 'Crystal Blocks of Yankee Coldness' to India and Brazil." *Revista espaco academico* 142 (2012): 162–77.

Heschong, Lisa. *Thermal Delight in Architecture*. Cambridge, MA: MIT Press, 1979.

Hess, Volker. "Standardizing Body Temperature: Quantification in Hospitals and
Daily Life, 1850–1900." In *Body Counts: Medical Quantification in Historical and
Sociological Perspectives*, edited by Gérard Jorland, Annick Opinel, and George
Weisz, 109–26. Montreal: McGill-Queen's University Press, 2005.

Heuer, Christopher P. *Into the White: The Renaissance Arctic and the End of the Image*.
Brooklyn: Zone, 2019.

Hobart, Hiʻilei Julia Kawehipuaakahaopulani. "Cooling the Tropics: Ice, Indigene-
ity, and Hawaiian Refreshment." Unpublished manuscript.

Hoberman, John M. *Black and Blue: The Origins and Consequences of Medical Racism*.
Berkeley: University of California Press, 2012.

Hogan, Mél, and Sarah T. Roberts. "Archives Melting (and Meltdowns)." In *After Ice:
Cold Humanities for a Warming Planet*, edited by Rafico Ruiz, Paula Schönach, and
Rob Shields. Unpublished manuscript.

hooks, bell. "Eating the Other: Desire and Resistance." In *Black Looks: Race and
Representation*, 21–39. Boston: South End Press, 1992.

Horikoshi, Satoshi, Robert F. Schiffmann, Jun Fukushima, and Nick Serpone.
Microwave Chemical and Materials Processing: A Tutorial. Singapore: Springer
Nature, 2018.

Horn, Eva. "The Aesthetics of Heat: For a Cultural History of Climate in the Age of
Global Warming." *Metaphora* 2 (2017): 1–16.

Hough, Theodore, and William T. Sedgwick. *The Human Mechanism: Its Physiology
and Hygiene and the Sanitation of Its Surroundings*. Boston: Ginn and Company,
1906.

Houghten, F. C., and C. P. Yagloglou. "Determination of the Comfort Zone." *ASHVE
Transactions* 29 (1923): 361–84.

Houghten, F. C., and C. P. Yagloglou. "Determining Equal Comfort Lines." *ASHVE
Transactions* 29 (1922): 165–76.

Howe, Bruce M., Brian K. Arbic, Jérome Aucan, Christopher R. Barnes, Nigel Bay-
liff, Nathan Becker, Rhett Butler, et al. "SMART Cables for Observing the Global

Ocean: Science and Implementation." *Frontiers in Marine Science* 6 (2019), article no. 424. http://doi.org/10.3389/fmars.2019.00424.

Howe, Cymene. "On Cryohuman Relations." In *After Ice: Cold Humanities for a Warming Planet*, edited by Rafico Ruiz, Paula Schönach, and Rob Shields. Unpublished manuscript.

Howe, Cymene, and Dominic Boyer. "Redistributions: From Atmospheric Carbon to Melting Cryospheres to the World Ocean." *e-flux Architecture*, September 21, 2018. https://www.e-flux.com/architecture/accumulation/212496/redistributions/.

Hughes, Rupert, dir. *Souls for Sale*. Los Angeles: Goldwyn Pictures, 1923.

Hulme, Mike. *Weathered: Cultures of Climate*. London: Sage, 2017.

Humphreys, M. A. "Thermal Comfort Temperatures and the Habits of Hobbits." In *Standards for Thermal Comfort: Indoor Air Temperature Standards for the 21st Century*, edited by Fergus Nicol, Michael Humphreys, Oliver Sykes, and Susan Roaf, 3–13. Abingdon, UK: Taylor and Francis, 1995.

Huntington, Ellsworth. "Influenza and the Weather in the United States in 1918." *Scientific Monthly*, November 1923, 462–71.

Hutchison, Phillip J. "Journalism and the Perfect Heat Wave: Assessing the Reportage of North America's Worst Heat Wave, July–August 1936." *American Journalism* 25, no. 1 (2008): 31–54.

Ingels, Margaret. *Willis H. Carrier: Father of Air-Conditioning*. Garden City, NY: Country Life Press, 1952.

Irving, L., and J. S. Hart. "The Metabolism and Insulation of Seals as Bare-Skinned Mammals in Cold Water." *Canadian Journal of Zoology* 35, no. 4 (1957): 497–511.

Jacobs, Meg. *Panic at the Pump: The Energy Crisis and the Transformation of American Politics in the 1970s*. New York: Hill and Wang, 2016.

Jennings, Eric. *Curing the Colonizers: Hydrotherapy, Climatology, and French Colonial Spas*. Durham, NC: Duke University Press, 2006.

Jiao, Yanmei, Yuanlong Li, Yonggang Wen, YewWah Wong, Kok Chuan Toh, Chee Cheng Chua, and Wilson Ang. "Thermal Analysis for Underground Data Centres in the Tropics." *Energy Procedia* 143 (2017): 223–29.

Johnston, Alan R., and Harold Kirkham. "An Electric Field Meter and Temperature Measurement Techniques for the Power Industry." *Proceedings of SPIE: Fiber Optic and Laser Sensors IV*, vol. 0718 (April 28, 1987): 134–41.

Jones, E. "The Demonstration of Collateral Venous Circulation in the Abdominal Wall by Means of Infra-red Photography." *American Journal of the Medical Sciences* 190, no. 4 (1935): 478–85.

Jousset, Philippe, Thomas Reinsch, Trond Ryberg, Hanna Blanck, Andy Clarke, Rufat Aghayev, Gylfi P. Hersir, Jan Henninges, Michael Weber, and Charlotte M. Krawczyk. "Dynamic Strain Determination Using Fibre-Optic Cables Allows Imaging of Seismological and Structural Features." *Nature Communications* 9 (2018), article no. 2509. https://doi.org/10.1038/s41467-018-04860-y.

Jue, Melody. *Wild Blue Media: Thinking through Seawater*. Durham, NC: Duke University Press, 2020.

Jue, Melody, and Rafico Ruiz. "Thinking with Saturation beyond Water: Thresholds, Phase Change, and the Precipitate." In *Saturation: An Elemental Politics*, edited by Melody Jue and Rafico Ruiz, 1–26. Durham, NC: Duke University Press, 2021.

Kantor, Bernard R. "Infrared Motion-Picture Technique in Observing Audience Reactions." *Journal of the Society of Motion-Picture and Television Engineers* 64 (1955): 626–28.

Kember, Sarah, and Joanna Zylinska. *Life after New Media: Mediation as a Vital Process*. Cambridge, MA: MIT Press, 2012.

Kempton, Willett, and Loren Lutzehiser. Introduction to "Devoted to Air Conditioning: The Interplay of Technology, Comfort, and Behavior" (special issue). *Energy and Buildings* 18, nos. 3–4 (1992): 171–76.

Kiesling, Barrett C. *Talking Pictures: How They Are Made, How to Appreciate Them*. Richmond, VA: Johnson Publishing Company, 1937.

Kilduffe, Rodert A. "Synthetic Fever as a Treatment for Disease." *American Journal of Nursing* 31, no. 10 (1931): 1125–28. http://doi.org/10.2307/3410355.

Kingma, Boris, and Wouter van Marken Lichtenbelt. "Energy Consumption in Buildings and Female Thermal Demand." *Nature Climate Change* 5, no. 12 (2015): 1054–56.

Kishore, Ravi Anant, Amin Nozariasbmarz, Bed Poudel, Mohan Sanghadasa, and Shashank Priya. "Ultra-High Performance Wearable Thermoelectric Coolers with Less Materials." *Nature Communications* 10 (2019), article no. 1765.

Klein, Sanford, and Gregory Nellis. *Thermodynamics*. Cambridge: Cambridge University Press, 2012.

Klinenberg, Eric. *Heat Wave: A Social Autopsy of Disaster in Chicago*. Chicago: University of Chicago Press, 2002.

Krause, Paul. *The Battle for Homestead, 1880–1892: Politics, Culture, and Steel*. Pittsburgh: University of Pittsburgh Press, 1992.

Lancet. "Influence of the Cerebellum on the Genitals." 2 (May 2, 1835): 168.

Langley, Samuel Pierpont. "The Bolometer and Radiant Energy." *Proceedings of the American Academy of Arts and Sciences* 16 (May 1880–June 1881): 342–58.

Lara, Ali. "Affect, Heat and Tacos: A Speculative Account of Thermoception." *Senses and Society* 10, no. 3 (2015): 275–97.

Lawson, Ray N., G. D. Wlodek, and D. R. Webster. "Thermographic Assessment of Burns and Frostbite." *Canadian Medical Association Journal* 84, no. 20 (1961): 1129–31.

Lee, Spike. *A Companion to the Universal Pictures Film "Do the Right Thing: A Spike Lee Joint."* New York: Fireside, 1989.

LeMenager, Stephanie. "Living with Fire (Hot Media)." *e-flux Architecture*, September 18, 2018. www.e-flux.com/architecture/accumulation/212491/living-with-fire-hot-media.

LeVine, Susan. *The Active Denial System: A Revolutionary, Nonlethal Weapon for Today's Battlefield*. Washington, DC: National Defense University Center for Technology and National Security Policy, June 2009.

Lévi-Strauss, Claude. *Mythologiques*. Vol. 1, *The Raw and the Cooked*. Translated by John and Doreen Weightman. 1964. Reprint, Chicago: University of Chicago Press, 1983.

Li, Tingye. Preface to *Optical Fiber Communications: Fiber Fabrication*, edited by Tingye Li, xi–xii. Orlando: Academic Press, 1985.

Lichtenstein, Alex. *Twice the Work of Free Labor: The Political Economy of Convict Labor in the New South*. London: Verso, 1996.

Lin, Jun, Weihao Huang, Muchen Wen, Dehong Li, Shuyi Ma, Jiawen Hua, Hang Hu, et al. "Containing the Spread of Coronavirus Disease 2019 (COVID-19): Meteorological Factors and Control Strategies." *Science of the Total Environment* 744 (2020), article no. 140935.

Lindsay, D. C. "Air Conditioning as Applied in Theaters and Film Laboratories." *Transactions of the Society of Motion Picture Engineers* 9, no. 30 (1927): 334–64.

Lindsey, Nathaniel J., T. Craig Dawe, and Jonathan B. Ajo-Franklin. "Illuminating Seafloor Faults and Ocean Dynamics with Dark Fiber Distributed Acoustic Sensing." *Science* 366, no. 6469 (2019): 1103–7. http://doi.org/10.1126/science.aay5881.

Liu, Alan. *The Laws of Cool: Knowledge Work and the Culture of Information*. Chicago: University of Chicago, 2004.

Lovecraft, H. P. "Cool Air." *Tales of Magic and Mystery*, March 1928.

Lovecraft, H. P. "At the Mountains of Madness." *Astounding Stories*, February 1936.

Lunstrum, Elizabeth. "Green Militarization: Anti-Poaching Efforts and the Spatial Contours of Kruger National Park." *Annals of the Association of American Geographers* 104, no. 4 (2014): 816–32.

MacKenzie, Scott, and Anna Westerståhl Stenport. *Films on Ice: Cinemas of the Arctic*. Edinburgh: Edinburgh University Press, 2014.

Mandal, Chandi C., and M. S. Panwar. "Can the Summer Temperatures Reduce COVID-19 Cases?" *Public Health* 185 (2020): 72–79.

Mann, Thomas. *The Magic Mountain*. 1924. Reprint, New York: Knopf, 1995.

Marks, Laura U. *The Skin of the Film: Intercultural Cinema, Embodiment, and the Senses*. Durham, NC: Duke University Press, 2000.

Marshall, S. J. *Infrared Thermography of Buildings: An Annotated Bibliography*. Hanover, NH: Army Corps of Engineers, 1977.

Martin, Niall. "As 'Index and Metaphor': Migration and the Thermal Imaginary in Richard Mosse's 'Incoming.'" *Culture Machine* 17 (2018): 1–19.

Marx, Karl, and Friedrich Engels. *Manifesto of the Communist Party*. 1848. Reprint, Chicago: Charles H. Kerr, 1910.

Massachusetts General Court. *Investigation into the Management and Discipline of the State Reform School at Westborough*. Boston: Albert J. Wright, 1877.

Massumi, Brian. *Ontopower: War, Powers, and the State of Perception*. Durham, NC: Duke University Press, 2015.

Mbembe, Achille. "Necropolitics." Translated by Libby Meintjes. *Public Culture* 15, no. 1 (2003): 11–40.

McIntyre, D. A. "Evaluation of Thermal Discomfort." In *Indoor Air: Proceedings of the Third International Conference on Indoor Air Quality and Climate*, edited by Birgitta Berglund, Thomas Lindvall, and Jan Sundell, 147–58. Stockholm, 1984.

McLuhan, Marshall. *Understanding Media: The Extensions of Man*. Cambridge, MA: MIT Press, 1994.

McPherson, Tara. "U.S. Operating Systems at Mid-Century: The Intertwining of Race and UNIX." In *Race after the Internet*, edited by Lisa Nakamura and Peter Chow-White, 21–27. New York: Routledge, 2012.

Middleton, W. E. Knowles. *A History of the Thermometer and Its Uses in Meteorology*. Baltimore: Johns Hopkins University Press, 1966.

Miles, Christopher. "The Combine Will Tell the Truth: On Precision Agriculture and Algorithmic Rationality." *Big Data and Society* 6, no. 1 (2019): 1–12.

Miller, Michael. *The Internet of Things: How Smart TVs, Smart Cars, Smart Homes, and Smart Cities Are Changing the World*. Indianapolis: Que, 2015.

Mims, Forrest M., III. "The First Century of Lightwave Communications." *International Fiber Optics and Communications Handbook and Buyers Guide* (1981–1982): 6–23.

Mogensen, Meryl F., and Horace B. English. "The Apparent Warmth of Colors." *American Journal of Psychology* 37, no. 3 (1926): 427–28.

Montesquieu, Charles Louis de Secondat. "Of Laws as Relative to the Nature of the Climate." In *The Complete Works of M. de Montesquieu*, 14:292–309. London: T. Evans, 1777.

Monthly Review of the US Bureau of Labor Statistics. "Report of New York State Commission on Ventilation." 2, no. 5 (1916): 48–51.

Mosse, Richard. "Transmigration of the Souls." In *Incoming*. London: Mack, 2017.

Mukherjee, Rahul. *Radiant Infrastructures: Media, Environment, and Cultures of Uncertainty*. Durham, NC: Duke University Press, 2020.

Mulvin, Dylan, and Jonathan Sterne. "Introduction: Temperature Is a Media Problem." *International Journal of Communication* 8 (2014): 2496–2503.

Mulvin, Dylan, and Jonathan Sterne, eds. "Media Hot and Cold" (special issue). *International Journal of Communication* 8 (2014).

Mumford, Lewis. *Technics and Civilization*. London: Routledge and Kegan Paul, 1955.

Nading, Alex. "Heat." Theorizing the Contemporary, *Fieldsights*, April 6, 2016. https://culanth.org/fieldsights/heat.

Nature. "Annual Exhibition of the Royal Photographic Society." 3281, no. 1301 (1932): 444.

Nature. "MM. Osty's Investigations of Rudi Schneider (From a Correspondent)." 133 (1934): 747–49.

Nietzsche, Friedrich. "The Wanderer and His Shadow." In *Human, All Too Human*, translated by R. J. Hollingdale, 301–96. Cambridge: Cambridge University Press, 1996.

Nixon, Rob. *Slow Violence and the Environmentalism of the Poor*. Cambridge, MA: Harvard University Press, 2011.

Noon, Derek. "Negotiating a Quantum Computation Network: Mechanics, Machines, Mindsets." PhD diss., Carleton University, 2016.

O'Connell, Allan F., James D. Nichols, and Ullas K. Karanth, eds. *Camera Traps in Animal Ecology: Methods and Analyses*. Tokyo: Springer Japan, 2011.

Ong, Boon Lay. "Introduction: Environmental Comfort and Beyond." In *Beyond Environmental Comfort*, edited by Boon Lay Ong, 1–16. London: Routledge, 2013.

Ong, Boon Lay. "Warming Up to Heat." *Senses and Society* 7, no. 1 (2012): 5–21.

Osman, Michael. *Modernism's Visible Hand: Architecture and Regulation in America*. Minneapolis: University of Minnesota Press, 2018.

Packard, Fred A. *Memorandum of a Late Visit to the Auburn Penitentiary*. Philadelphia: J. Harding, 1842.

Parks, Lisa. "Vertical Mediation and the U.S. Drone War in the Horn of Africa." In *Life in the Age of Drone Warfare*, edited by Lisa Parks and Caren Kaplan, 134–57. Durham, NC: Duke University Press, 2017.

Parr, Joy. *Sensing Changes: Technologies, Environments, and the Everyday, 1953–2003*. Vancouver: UBC Press, 2010.

Parsons, Ken. *Human Thermal Environments: The Effects of Hot, Moderate, and Cold Environments on Human Health, Comfort and Performance*. 2nd ed. London: Taylor and Francis, 2003.

Parsons, Ken C. Introduction to *Standards for Thermal Comfort: Indoor Air Temperature Standards for the 21st Century*, edited by Fergus Nicol, Michael Humphreys, Oliver Sykes, and Susan Roaf, xiii–xiv. Oxford: Taylor and Francis, 1995.

Pergravis. "Is Cooling the Top Efficiency Metric for Tropical Data Centers?" April 26, 2016. http://pergravis.com/industry-insights/blog/is-cooling-the-top -efficiency-metric-for-tropical-data-centers/.

Perkinson, Robert. "'Hell Exploded': Prisoner Music and Memoir and the Fall of Convict Leasing in Texas." *Prison Journal* 89, no. 1 (2009): 54–69.

Peters, John Durham. *The Marvelous Clouds: Toward a Philosophy of Elemental Media*. Chicago: University of Chicago Press, 2015.

Peterson, Marina. "Sensory Attunements: Working with the Past in the Little Cities of Black Diamonds." *South Atlantic Quarterly* 115, no. 1 (2016): 89–111.

Pollack, Henry N. *A World without Ice*. New York: Avery, 2010.

Potter, Caroline. "Sense of Motion, Senses of Self: Becoming a Dancer." *Ethnos: Journal of Anthropology* 73, no. 4 (2008): 444–65.

Potter, Ned. "Facebook Plans Server Farm in Sweden: Cold Is Great for Servers." *ABC News*, October 27, 2011. http://abcnews.go.com/Technology/facebook-plans -server-farm-arctic-circle-sweden/story?id=14826663.

Povinelli, Elizabeth A. *Geontologies: A Requiem to Late Liberalism*. Durham, NC: Duke University Press, 2016.

Prescott, George Bartlett. *Bell's Electric Speaking Telephone: Its Invention, Construction, Application, Modification, and History*. New York: D. Appleton, 1884.

Purdon, Henry Samuel. "On the Use of Ice in Hysteria." *Medical Circular* 26, November 29, 1865.

Radin, Joanna. *Life on Ice: A History of New Uses for Cold Blood*. Chicago: University of Chicago Press, 2017.

Radin, Joanna, and Emma Kowal. "Introduction: The Politics of Low Temperature." In *Cryopolitics: Frozen Life in a Melting World*, edited by Joanna Radin and Emma Kowal, 3–26. Cambridge, MA: MIT Press, 2017.

Radin, Joanna, and Emma Kowal, eds. *Cryopolitics: Frozen Life in a Melting World.* Cambridge, MA: MIT Press, 2017.

Rankine, A. O. "The Transmission of Speech by Light." *Nature* 111 (1923): 744–45.

Rao, M. Mukunda. *Optical Communication.* Hyderabad, India: Universities Press, 2000.

Rawling, S. O. *Infra-red Photography.* 3rd ed. London: Blackie and Son, 1939.

Razack, Sherene. *Dying from Improvement: Inquests and Inquiries into Indigenous Deaths in Custody.* Toronto: University of Toronto Press, 2015.

Rees, Jonathan. *Refrigeration Nation: A History of Ice, Appliances, and Enterprise in America.* Baltimore: Johns Hopkins University Press, 2013.

Rejali, Darius. *Torture and Democracy.* Princeton, NJ: Princeton University Press, 2007.

Ritchie, Michael. *Please Stand By: A Prehistory of Television.* Woodstock, NY: Overlook Press, 1994.

Roaf, Sue, Luisa Brotas, and Fergus Nicol. "Counting the Costs of Comfort." *Building Research and Information* 43, no. 3 (2015): 269–73.

Roberts, Donald M. "Icebergs as a Heat Sink for Power Generation." In *Iceberg Utilization: Proceedings of the International Conference Held at Ames, Iowa,* edited by A. A. Husseiny, 674–89. New York: Pergamon, 1978.

Rodriguez-Palacios, Alex, Mathew Conger, and Fabio Cominelli. "Nonmedical Masks in Public for Respiratory Pandemics: Droplet Retention by Two-Layer Textile Barrier Fully Protects Germ-Free Mice from Bacteria in Droplets." *bioRxiv,* April 6, 2020. https://doi.org/10.1101/2020.04.06.028688.

Rogers, A. J. "Invited Paper Distributed Sensors: A Review." *Proceedings of SPIE: Fiber Optic Sensors II,* vol. 0798 (October 14, 1987): 26–35. https://www.spiedigitallibrary.org/conference-proceedings-of-spie/0798/0000/Invited-Paper-Distributed-Sensors-A-Review/10.1117/12.941081.short?SSO=1.

Rohles, Frederick H., Jr. "Temperature and Temperament: A Psychologist Looks at Comfort." *ASHRAE Journal* 49 (2007): 14–22.

Ronchese, F. "Infra-red Photography in the Diagnosis of Vascular Tumors." *American Journal of Surgery* 37, no. 3 (1937): 475–77.

Rossi, R. M. "High-Performance Sportswear." In *High-Performance Apparel: Materials, Development, and Applications,* edited by John McLoughlin and Tasneem Sabir, 341–56. Duxford, UK: Woodhead Publishing, 2018.

Royston, Sarah. "Dragon-Breath and Snow-Melt: Know-How, Experience and Heat Flows in the Home." *Energy Research and Social Science* 2 (2014): 148–58.

Rubio, Fernando Domínguez. "Preserving the Unpreservable: Docile and Unruly Objects at MoMA." *Theory and Society* 43, no. 6 (2014): 617–45.

Rubio, Fernando Domínguez. *Still Life: Ecologies of the Modern Imagination at the Art Museum.* Chicago: University of Chicago Press, 2020.

Ruiz, Rafico. "Grenfell Cloth." In *New Materials: Towards a History of Consistency,* edited by Amy E. Slaton, 237–70. Amherst, MA: Lever, 2020.

Ruiz, Rafico. "Phase State Earth: Ice at the Ends of Climate Change." Unpublished manuscript.

Ruiz, Rafico, Paula Schönach, and Rob Shields. "Introduction: Experiencing after Ice." In *After Ice: Cold Humanities for a Warming Planet*, edited by Rafico Ruiz, Paula Schönach, and Rob Shields. Unpublished manuscript.

Santana, Dora Silva. "Trans* Stellar Knot-Works: Afro Diasporic Technologies, Transtopias and Accessible Futures." Unpublished manuscript.

Sargent, Epes Winthrop. *Picture Theater Advertising*. New York: Chalmers Publishing Company, 1915.

Schaefer, Eric. *"Bold! Daring! Shocking! True!": A History of Exploitation Films, 1919–1959*. Durham, NC: Duke University Press, 1999.

Schereschewsky, J. W. "Heating Effect of Very High Frequency Condenser Fields on Organic Fluids and Tissues." *Public Health Reports* 48, no. 29 (1933): 844–58. http://doi.org/10.2307/4580855.

Schönach, Paula. "Natural Ice and the Emerging Cryopolis: A Historical Perspective on Urban Cold Infrastructure." *Culture Machine* 17 (2018): 1–25.

Schüll, Natasha Dow. "Data for Life: Wearable Technology and the Design of Self-Care." *BioSocieties* 11, no. 3 (2016): 317–33.

Science News-Letter. "Radio Produces Artificial Fever." 14, no. 392 (1928): 227.

Seemann, J., Y. I. Chirkov, J. Lomas, and B. Primault. *Agrometeorology*. Berlin: Springer-Verlag, 1979.

Seligman, C. G. "Infra-red Photographs of Racial Types." *Nature* 133, no. 3356 (1934): 279–80.

Serres, Michel. "The Origin of Language: Biology, Information Theory, and Thermodynamics." In *Hermes: Literature, Science, Philosophy*, edited by David F. Bell, Michel Serres, and Josué V. Harari, 71–83. Baltimore: Johns Hopkins University Press, 1982.

Shachtman, Tom. *Absolute Zero and the Conquest of Cold*. New York: Houghton Mifflin, 2000.

Shakur, Assata. "Women in Prison: How We Are." *Black Scholar* 9, no. 7 (1978): 8–15.

Sharpe, Christina. *In the Wake: On Blackness and Being*. Durham, NC: Duke University Press, 2016.

Showalter, Elaine. "Hysteria, Feminism, and Gender." In *Hysteria Beyond Freud*, edited by Sander L. Gilman, Helen King, Roy Porter, G. S. Rousseau, and Elaine Showalter, 287–335. Berkeley: University of California Press, 1993.

Sifakis, Carl. "Sweatboxes." In *The Encyclopedia of American Prisons*, 251–52. New York: Facts on File, 2003.

Simmons, Kristen. "Settler Atmospherics." Theorizing the Contemporary, *Fieldsights*, November 20, 2017. https://culanth.org/fieldsights/settler-atmospherics.

Simonds, A. H., and L. H. Polderman. "Air Conditioning in Film Laboratories." *Journal of the Society of Motion Picture Engineers* 17, no. 5 (October 1931): 604–22.

Sloterdijk, Peter. *Terror from the Air*. Translated by Amy Patton and Steve Corcoran. Los Angeles: Semiotext(e), 2009.

Smith, Jen Rose. "Ice Core Coloniality: Meta-Data and Meta-Narratives." In *After Ice: Cold Humanities for a Warming Planet*, edited by Rafico Ruiz, Paula Schönach, and Rob Shields. Unpublished manuscript.

Smith, Jen Rose. "'Exceeding Beringia': Upending Universal Human Events and Wayward Transits in Arctic Spaces." *Environment and Planning D: Society and Space* 39, no. 1 (2021): 158–175.

Smith, Willoughby. "The Action of Light on Selenium." *Journal of the Society of Telegraph Engineers* 2 (1873): 31–33.

Smolen, James J., and Alex van der Spek. *Distributed Temperature Sensing: A DTS Primer for Oil and Gas Production*. The Hague, Netherlands: Shell International Exploration and Production B.V., 2003.

Soler, Lena, Frederic Wieber, Catherine Allamel-Raffin, Jean-Luc Gangloff, Catherine Dufour, and Emiliano Trizio. "Calibration: A Conceptual Framework Applied to Scientific Practices Which Investigate Natural Phenomena by Means of Standardized Instruments." *Journal for General Philosophy of Science* 44 (2013): 263–317.

Szczepaniak-Gillece, Jocelyn. *The Optical Vacuum: Spectatorship and Modernized American Theater Architecture*. Oxford: Oxford University Press, 2018.

Tawil-Souri, Helga. "Spectrum." Theorizing the Contemporary, *Fieldsights*, October 24, 2017. https://culanth.org/fieldsights/spectrum.

Taylor, Timothy D. *The Sounds of Capitalism: Advertising, Music, and the Conquest of Culture*. Chicago: University of Chicago Press, 2012.

Thévenot, Roger. *History of Refrigeration throughout the World*. Paris: International Institute of Refrigeration, 1979.

Thompson, Darla. "Circuits of Containment: Iron Collars, Incarceration and the Infrastructure of Slavery." PhD diss., Cornell University, 2014.

Thompson, Kristin, and David Bordwell. *Film History: An Introduction*. 2nd ed. New York: McGraw Hill Higher Education, 2003.

Tobías, Aurelio, and Tomás Molina. "Is Temperature Reducing the Transmission of COVID-19?" *Environmental Research* 186 (2020): 1–2.

Tredgold, Thomas. *Principles of Warming and Ventilating Public Buildings*. London: J. Taylor, 1824.

Troup, W. Annandale. *Therapeutic Uses of Infra-Red Rays*. London: Actinic, 1930.

Tuck, Eve. "Suspending Damage: A Letter to Communities." *Harvard Educational Review* 79 no. 3 (2009): 409–27.

Tyler, S. W., D. M. Holland, V. Zagorodnov, A. A. Stern, C. Sladek, S. Kobs, S. White, F. Suárez, and J. Bryenton. "Using Distributed Temperature Sensors to Monitor an Antarctic Ice Shelf and Sub–Ice-Shelf Cavity." *Journal of Glaciology* 59, no. 215 (2013): 583–91. http://doi.org/10.3189/2013JoG12J207.

van Dijck, José. *The Transparent Body: A Cultural Analysis of Medical Imaging*. Seattle: University of Washington Press, 2005.

Vannini, Phillip, and Jonathan Taggart. "Making Sense of Domestic Warmth: Affect, Involvement, and Thermoception in Off-Grid Homes." *Body and Society* 20, no. 1 (2014): 61–84.

Vassallo, Susi. *Report on the Risks of Heat-Related Illness and Access to Medical Care for Death Row Inmates Confined to Unit 32, Mississippi State Penitentiary, Parchman, Mississippi*. National Prison Project, American Civil Liberties Union. September 2002. https://www.aclu.org/files/pdfs/prison/vassallo_report.pdf.

Vaughan, Hunter. *Hollywood's Dirtiest Secret: The Hidden Environmental Cost of the Movies*. New York: Columbia University Press, 2019.

Velkova, Julia. "Data Centers as Impermanent Infrastructures." *Culture Machine* 18 (2019): 1–11.

Velkova, Julia. "Data That Warms: Waste Heat, Infrastructural Convergence and the Computation Traffic Commodity." *Big Data and Society* 3, no. 2 (2016): 1–10.

Venkat, Bharat Jayram. "Toward an Anthropology of Heat." *Anthropology News*, March 12, 2020. https://www.anthropology-news.org/index.php/2020/03/12/toward-an-anthropology-of-heat/.

Virilio, Paul. *War and Cinema: The Logistics of Perception*. Translated by Patrick Camiller. London: Verso, 2000.

Waldrep, Christopher. *Jury Discrimination: The Supreme Court, Public Opinion, and a Grassroots Fight for Racial Equality in Mississippi*. Athens: University of Georgia Press, 2010.

Walker, Janet. "Afterword: Climate Change as 'Matter out of Phase.'" In *Saturation: An Elemental Politics*, edited by Melody Jue and Rafico Ruiz, 306–11. Durham, NC: Duke University Press, 2021.

Wang, Zhe, Maohui Luo, Hui Zhang, Yingdong He, Ling Jin, Edward Arens, and Shichao Liu. "The Effect of a Low-Energy Wearable Thermal Device on Human Comfort." The 15th Conference of the International Society of Indoor Air Quality and Climate (ISIAQ), Philadelphia. July 1–9, 2018. https://escholarship.org/uc/item/5f2876gr.

Watt-Cloutier, Sheila. *The Right to Be Cold: One Woman's Fight to Protect the Arctic and Save the Planet from Climate Change*. Minneapolis: University of Minnesota Press, 2018.

Welch, Michael. "Chain Gangs." In *Encyclopedia of Prisons and Correctional Facilities*, vol. 1, edited by Mary Bosworth, 108–9. Thousand Oaks, CA: Sage, 2005.

Wells, H. G. *The War of the Worlds*. Leipzig: Bernhard Tauchnitz, 1898.

Wernimont, Jacqueline. *Numbered Lives: Life and Death in Quantum Media*. Cambridge, MA: MIT Press, 2018.

Whyte, Kyle Powys. "Indigenous Science (Fiction) for the Anthropocene: Ancestral Dystopias and Fantasies of Climate Change Crises." *Environment and Planning E: Nature and Space* 1, nos. 1–2 (2018): 224–42.

Wiener, Norbert. *Cybernetics, or Control and Communication in the Animal and the Machine*. 2nd ed. Cambridge, MA: MIT Press, 1965.

Willoughby, Christopher D. "'His Native, Hot Country': Racial Science and Environment in Antebellum American Medical Thought." *Journal of the History of Medicine and Allied Sciences* 72, no 3 (2017): 328–51.

Winter, Margaret, and Stephen F. Hanlon. "Parchman Farm Blues: Pushing for Prison Reforms at Mississippi State Penitentiary." *Litigation* 35, no. 1 (2008): 1–8.

Wishart, Paul B. "Honeywell: A Growth Company." *Analysts Journal* 12, no. 3. (1956): 125–28.

Witze, Alexandra. "Global 5G Wireless Deal Threatens Weather Forecasts." *Nature* 575, no. 7784 (2019): 577.

Wollen, Peter. "Fire and Ice." In *The Photography Reader*, edited by Liz Wells, 108–13. London: Routledge, 2003.

Woloshyn, Tania Anne. *Soaking Up the Rays: Light Therapy and Visual Culture in Britain, 1890–1940*. Manchester: Manchester University Press, 2017.

Wood, Robert W. "Photography by Invisible Rays." *Photographic Journal*, October 1910.

Woods, Rebecca J. H. *The Herds Shot Round the World: Native Breeds and the British Empire, 1800–1900*. Chapel Hill: University of North Carolina Press, 2017.

Yusoff, Kathryn. *A Billion Black Anthropocenes or None*. Minneapolis: University of Minnesota Press, 2018.

Yusoff, Kathryn. "Epochal Aesthetics: Affectual Infrastructures of the Anthropocene." *e-flux Architecture*, March 29, 2017. http://www.e-flux.com/architecture/accumulation/121847/epochal-aesthetics-affectual-infrastructures-of-the-anthropocene/.

Ziat, Mounia, Carrie-Anne Balcer, Andrew Shirtz, and Taylor Rolison. "A Century Later, the Hue-Heat Hypothesis: Does Color Truly Affect Temperature Perception?" In *Haptics: Perception, Devices, Control, and Applications*, edited by Fernando Bello, Hiroyuki Kajimoto, and Yon Visell, 273–80. Cham, Switzerland: Springer International, 2016.

Page numbers in italics refer to figures.

absolute zero, 191, 192
acclimatization, 11, 61, 93, 121, 241n52
Ackermann, Marsha, 33, 75
Adams, Sean Patrick, 37
adaptive-comfort models, 59–60, 65, 70, 71
advertisements: furnace regulators, 40, *41*;
    gramophone, 206; NBC, 203, *204*; radio,
    92–93, *94*; snow films, 79, 82, 84, 86;
    targeted to women, 40–42; thermostat,
    12, 34, 42–45, *43*, *44*, 46, *47*
aerial surveys, 176, 184
Afghanistan, 135–36, 187
AGA, 177, *178*, 179
Agbemabiese, Lawrence, 59
agency, 8, 36, 64; climate change and, 71;
    thermostat's, 42, 66, 68
agriculture, xiii, 10, 122, 250n15; broadcast
    and precision, 169, 179–82, 184; chicken
    farming, xiv; infrared imaging and, *173*, 174
air, medium of, 100, 101
air conditioners, 1, 6, 17, 21, 53, 60, 75; as a
    cold attraction, 76, 100, 108; for digital
    infrastructures, 209; engineers, 101–2;
    film manufacturing and, 200–202; home
    units, 103; installation in cinemas, 97,
    99–100, 103; for preservation of media,
    24, 194, 218; print industry and, 199, 208,
    250n26; in prisons, 124–25, 127, 130, 131;
    as a right, 129; studio productions and,
    202–3; on trains, 102–3; used for physical
    harm, 110–11, 125; wearable, 2, 22, 104,
    106; worker productivity and, 31–32,
    64–65
Alaimo, Stacy, 14, 130–31
Alaska, 81, 82, 83, 87

algorithms, 24, 63, 66, 68–69, 182, 212
alliesthesia, 107
Amazon, 61, 211
American Civil Liberties Union (ACLU),
    125, 136
American Society of Heating, Refrigerat-
    ing, and Air-Conditioning Engineers
    (ASHRAE), 211, 214; Standard 55, 56–60,
    70
American Society of Heating and Ventilat-
    ing Engineers (ASH&VE), 52, 53–54, 56–57
American South, 116, 118–19, 120
Ammons, Edward, 120
Amundsen, Roald, 80
analog *vs.* digital media, 208–9
Antarctic, 78, 97; expeditions, 75, 80–81,
    81, 103–4, 235n17; McMurdo Ice Shelf,
    161, *162*
anthrax, 219
antipoaching campaigns, *183*, 184
Antunes, Luis, 247n1
apartheid infrastructures, 169, 184
Apple iPod, 62
apps, 1, 32, 62, 64–65
architecture, 8, 14, 177, 191, 207, 226n17;
    normativity and uniformity, 33, 229n10;
    of sweatboxes, 115–18, 122, 131; thermal
    violence and, 124, 125, 128
archives, 206–7, 209, 210, 218
Arctic, 78, 95, 104, 209; climate change
    and, 219–20; colonization and, 5, 75, 76;
    expeditions, 75, 80, 92–93, 96, 235n17;
    Inuit culture, 27; portrayed in films, 21,
    25, 81–83, 87; tourism, 21
Aristotle, 10

artificial intelligence, 182, 216; autonomous vehicles, 24, 159, 161; enabled thermostats, 26, 61–69

atmosphere, 78, 79–80, 99, 107–8, 165; control of, 20, 52, 62, 70, 71, 195, 200–201; creation of, 83–84, 86–87, 104; fear of, 95; warmth as, 4

"At the Mountains of Madness" (Lovecraft), 97, 98

audiovisual exposures, 99–100

autonomy. See thermal autonomy

Bachelard, Gaston, 10, 73, 107

Baird, John Logie, 137, 146–49, 155, 165, 244n29

Baker, Thorne, 100

Barad, Karen, 18

Barber, Daniel, 8, 15, 33, 48, 56, 70, 229n10

bare life, 170, 185, 188

Bell, Alexander Graham, 23, 144–45, 149, 155

Bell Labs, 156–57

Beregow, Elena, 225n4

Berlant, Lauren, 4

biopolitics, 184, 185, 222, 226n15

biopower, 7, 226n15

Black people: Anti-Blackness, 129; autoworkers, 13, 227n48; racist beliefs about, 121; sweatboxing, 116–21, 127–28; women in prison, 127

Blum, Hester, 75, 88

bodies, 59, 222; Black people's, 121; calibration of, 17–18, 36–37, 70, 71, 104, 114; cooling of, 78, 96–97, 100, 104–7; heat emissions or exchanges of, 3–4, 15, 16, 105, 128, 139–40, 153; heat images of, 7, 173, 174, 185–88; hot/cold exposure and, 14, 35, 72–73, 109–11, 126–27, 129–31; managing of, 26, 170, 190; as media, 165, 175, 223; radio heat and, 150–53; sweatboxing and, 115–17, 121; thermal vision and, 166–68

bolometer, 171

Boyer, Dominic, 220, 226n15

Brager, Gail Schiller, 59–60

broadcast heat, 152

broadcast temperature systems, 32–34, 39, 56. See also central heating

buildings: cooling of, 5, 31–32; energy audits, 176; fluctuating temperatures and,

4; heating of, 6, 46; insulation, 142, 169, 176; temperature control, 50, 52, 64, 65, 231n69, 234n137; thermal comfort and, 56, 189; ventilation systems, 34, 53. See also windows

Burns, Edward N., 204–5

Butz, Alfred, 38

Byrd, Richard, 80, 93, 103

calibration: bodily, 17–18, 22, 36–37, 70, 71, 104, 114; of infrared cameras, 188; of media, 194; of personalized temperature, 131; of thermal subjects, 165, 168

calorific rays, 143

capitalism. See thermal capitalism

Carrier Corporation, 64–65, 75, 100, 199, 200

Carrier, Willis, 99, 199, 232n93

Carter, Jimmy, 50–51

Case, Theodore Willard, 145–46

central heating, 20, 33, 38–39, 64, 73, 152

Chronotherm, 42, 43

Chun, Wendy Hui Kyong, 27, 160

CIA, 126

cinemas, 167, 173, 247n1; cooling technologies, 95–97, 98, 99–104, 194; exhibitors and displays, 84, 86–91, 87, 89, 90; popularity, 78–79; projection booths, 202; racial labor, 91–92

civilization: alien, 97; development of, 10–11, 33, 74, 250n15; Western, 25, 195, 210

civil rights, 48, 125

Clicquot Club Eskimos, The (1923), 80, 93–94

climate change, 1–3, 36, 75, 78, 121, 130, 223; academic research on, 221–22; agency and, 71; distributed temperature sensing and, 162–63; environmental determinism and, 10; global internet and, 139; meltings and solidifications, 219–21; mitigating, 21, 128, 189, 224; radiant thermal media and, 165; recognition of, 19; smart thermostats and, 12–13, 26, 62–64; temperature fluctuations and, 210, 212, 218; thermal violence and, 22, 27, 114, 122, 124, 126, 131–32, 222; thermal vision and, 189–90; threat of, 216, 218

climate(s): Anti-Blackness and, 129; artificial, 53, 99; California, 201; cold,

11, 195–97, 216; control, 61–62, 201, 207; COVID-19 pandemic and, 26; data centers and, 213–14, *215*, 216; determinism, 11, 75, 121, 195; extreme, 113; microclimates, 62, 153, 164; Other, 107; tropical, 59, 74, 210, 213–14, 216, 218, 219; war and, 79–80; world leadership and, 195, *196*, 250n15

clothing, 154, 164; cooling, 22, 104, 106; emissivity of, 142, 244n9; thermal insulation of, 6, 57–58, 244n29

coal, 10, 38–39, 76, 201, 203

cold attractions, 75–77, 91, 103–4, 105, 107–8

coldsploitation media, 152, 157, 167; cinematic cooling systems, 95–97, 99–104; fictional stories, 97, 219; films and exhibitors, 80–84, *85*, 86–91, 236n36; meaning, 21, 76–77; radio programs, 21, 80, 92–94, *94*; stores and displays, 91

colonization, 5, 8, 21, 94, 223; acclimatization and, 11, 93, 241n52; climate change and, 222; coldsploitation films and, 87–88; cooling technologies and, 74–75, 96

color, 86, 148; blind, 170, 187; lights and, 143, 190; printing, 194, 197–200; temperature and, 2, 167–68, 190, 248n27

comfort zone chart, 70, 152, 232n93; air-conditioning and, 100, 101, 102; ASH&VE study, 34, 53–54, *55*, 56, 232n90. *See also* thermal comfort

Comfy app, 65

commodities, 74, 76, 95, 207, 210

computers, 19, 50, 209, 210; quantum, 191–92, *193*. *See also* data centers

conduction, 15, 78, 144, 209

conductive media, 21, 23, 26, 78, 104–5, 142; description, 15–16; digital technologies, 77, 107

Congo, 5

conservation, 179, 189; energy, 50, 63, 176; wildlife, 169, 182–85, *183*

convective media, 20, 22, 23, 38, 77, 142; air conditioners as, 99, 199; coldsploitation as, 21, 78, 91, 104; description, 15; thermal violence and, 112; thermocultural shift to, 76

"Cool Air" (Lovecraft), 96–97

cooling systems, 33, 56, 62; air deflectors, 124; in cinemas, 95–97, *98*, 99–104; cold

storage, 207; colonial projects and, 74–75; desire for, 73–74; digital infrastructures and, 192, 208–10; industrial, 64–65; personalized, 22, 65, 68, 77–78, 104–5; for preservation of media, 194–95, 207–8, 217; on trains, 102–3; tropical areas and, 59, 213–14. *See also* air conditioners

Cooper, Gail, 53, 64–65, 99

corporeal punishment, 116–18, 122, 124–25, 241n38

COVID-19 pandemic, 24, 26, 169

Cowen, Ruth Schwartz, 41

critical temperature studies, 8, 28, 126

CrowdComfort, 64

crowd control, 135–36

cryopolitics, 8, 110, 195, 207–8, 218

cybernetics, 212, 225n5, 253n88

cybersecurity, 210

data centers, 25, 192, 209–14, *215*, 216

data sharing, 63, 66

Day, Deanna, 7

de Dear, Richard J., 59–60

Deleuze, Gilles, 20

democracy, 65, 69

Dibble, Samuel Edward, 152

digital systems, 1, 3, 24, 57; climate change and, 19, 26–27; computing, 191–92, *193*; cooling of, 208–9; environmental costs of, 77, 194, 224; fiber-optic infrastructure and, 156, 160–61; smart thermostats, 21, 50, 61–69; thermal autonomy and, 70–71, 131; thermoelectric, 106–7. *See also* data centers

Dillon, Stephen, 127

domestic care, 21, 34, 36–37, 41, 45

Do the Right Thing (1989), 167

Douglas, Mary, 5

drones, 4, 179, 181–82, 184, *185*, 187

Eastman Kodak, 146, 171–72, 177

Ecobee, 12, 61

electricity, 105–6, 144, 211, 241n50; grids, 9, 67, 213

electromagnetic radiation: description of, 15–16; emissions, 139–40, 165, 175–76; first form of, 143; human body and, 135, 153; spectrum, 137, 140, *141*, 152, 161, 175

Embr Wave, 22, 106–7, 142

emissivity, 175–76; in agricultural production, 180–81; of bodies, 140–41, 153, 223; of minerals, 179; of surfaces and substances, 140–42, 247n27

energopower, 7, 226n15

energy: audits, 176, 177, 183, 185; companies, 67–68; conservation, 50, 63, 67; crisis (1970s), 50–51, 169, 176, 189–90; data centers' consumption of, 210, 213–14; power usage efficiency (PUE), 214; renewable, 216

EnergyHub, 67

Engels, Friedrich, 220

entanglement. *See* thermal entanglement

environmental determinism, 10–14, 25, 26, 33, 56, 195

environmental monitoring, 162–64, 175, 188, 189; of wildlife, 169, 182–85, *183*

epigenetic research, 14

Equiano, Olaudah, 127–28

ethnography, 8, 192, 226n24

Everest, Mount, 80

expeditions. *See* polar explorations

exploitation: atmospheric, 91, 95–96; cold and cooling, 95, 100, 102; films, 79, 82–84, 86–89, *87*, *89*, 95; radio, 94

extensions of media, 217–18

extractive industries, 162, 179, *180*

Facebook, 31, 209, 216

Fadell, Tony, 62, 63

family unit, 41, 43–45, 46, 49, 52, 54

Fanger, Povl Ole, 57–59, 64

fans, 124, 130, 131, 190, 209; in cinemas, 79, 89, 95–96, *96*, 101

farming. *See* agriculture

Fennell, Catherine, 6

fever and fever machines, 1, 23, 137, 149–54, 166

fiber-optic cables: distributed temperature sensing, 139, 161–64, 165; early experiments, 156–60, *158*, *159*; internet data transmission, 1, 17, 23; visible light transmission, *160*, 160–61

Filament Mind (Brush and Lee), *160*

film manufacturing, 199–202, 206

films. *See* snow films

fireplaces, 36, 39–40

Flaherty, Robert, 81, 88, 93

FLIR Systems, 169, 177, 179, 189; for wildlife monitoring, 169, 182–85, *183*

Florida State Prison. *See* Union Correctional Institution

food: consumption, 74, 235n16; cooking, 10, 14, 153, 166; organic, 182; restricting or withholding, 22, 110, 116, 117–18; security, 179, 181; systems, 74, 122, 222

Ford Motor Company, 13, 227n48

Fort Lee (NJ), 200, 201

fresh air advocates, 53

Frozen North, The (1922), 82

furnaces: dampers, 37–38; draft regulators, 40, *41*, 44; hot-air, 38–39

Furuhata, Yuriko, 15, 20, 62, 77

Gabrys, Jennifer, 168

gender, 5, 7, 39, 113–14, 225n7; comfort zones and, 54, 56; labor and, 37, 38, 45; thermostat wars, 32, 35, 43–44, *44*, 48–52, 49, 167–68

General Electric Research Laboratory, 149–50, 153

geological sensing, 163–64

geopolitics, 27, 87, 195, 197

GilFillan, S. Colum, 75, 195–96, 250n15

Gilmore, Ruth Wilson, 113

Gitelman, Lisa, 226n11

glass: fibers, 156–57, 160, 161, 246n77; thermal fracture of, 4

global warming, 2, 12–13, 106; melting of permafrost, 210, 219–20. *See also* climate change

Google, 182, 211, 216

Gordon, Lewis, 119

Green Grid, 214

Guattari, Félix, 20

Gumbs, Alexis Pauline, 13

Haapoja, Terike, 187

hackers, 68–69, 210

haptic visuality, 21, 77, 90, 167, 186, 189

Hawai'i, 5, 74, 93, *94*

heat exchanges, 5, 7, 10, 115; bodies and, 3–4, 15, 16, 121, 187, 225n4; conduction, 15, 209; data centers and, 210–11; of digital systems, 192, 194

heat images. *See* infrared cameras and imaging
Heat Maps (Mosse), 170, 185–87, *186*
heat rays, 23–24, 138, 154, 164, 168; guns, 1, 136–37, 155; radio, 149–53; telegraph, 17, 23, 24; transmitters, 146; Wells's description of, 136, 145
heat regulators, 40–42. *See also* Minneapolis-Honeywell Regulator Company
Hell's Highway (1932), 119
Henderson, Sákéj, 109
Herschel, William, 143, 148, 154
Heschong, Lisa, 18, 73, 107
Hive thermostats, 66, 67, 68–69
Hobart, Hi'ilei Julia Kawehipuaakahaopulani, 5, 74
home heating systems, 9, 25, 37–42, 56, 229n14; zone control, 46, 48
Honeywell. *See* Minneapolis-Honeywell Regulator Company
hooks, bell, 104
Horn, Eva, 8, 12
Hosmer, Helen R., 150
hot and cold metaphors, 9–12, 170
Howe, Cymene, 78, 220
hue-heat research, 167, 176, 189–80
Hulme, Mike, 8
humidity, 53–54, 126, 195, 200, 202, 213; print industry and, 197, 199, 208; relative, 57, 101
humoral medical theory, 10, 56
Huntington, Ellsworth, 10, 26, 75
hypothermia, 79, 109, 111

IBM, 191–92, *193*
ice, 110, 250n12; attractions, 91; climate change and melting, 162–63, 219–21; core samples, 75, 207; in films and exhibitions, 82–83, 86–90; geographies, 75; hotels, 72; production and distribution of, 73–75, 76, 93–94; sheets, 26, 161; thermal emissivity of, 141, 248n27; vests, 22, 104; wristbands, 15, 106–7
iceboxes (or *hieleras*), 76, 110–11, 112, 164
ice cream, 74, 87, 91, 94, 99, 105
identity, 75, 207
image transmission, 146–47

imaging. *See* infrared cameras and imaging
Indigenous peoples, 74, 109, 222; racist depictions of, 75, 87–88, 91, 104
influenza pandemic (1918), 80, 149
information theory, 9–10, 16
infrared cameras and imaging, 6, 24, 26, 68, 147, 168–70, 222; AGA Thermovision Model, 177, *178*; agriculture management and, 179–82; commercial uses, 188–89; energy conservation and, 176; medical research and, 173–74, *174*; mining and, 179, *180*; Mosse's exhibition, 185–87, *186*; photography and plates, 171–73, *172*, *173*; temperature and, 175–76; wildlife management and, 182–85, *183*
infrared radiation: early media experiments, 143–48, *147*, 164; emissions, 15, 140–42, 171, 175–76, 247n27; invisibility of, 160–61; limitations of, 155; military and medical research and, 148–49; spectrum, 137–39, 157, 159, 165; wavelengths and fiber optics, 1, 23, 139, 157–60, *158*, *159*
Intel, 211
internet, 27, 62, 73, 160, 192, 197; infrared waves and, 1, 23, 26, 155; infrastructure, 209, 213, 216, 217; as a thermometer, 24, 139, 163–64, 165
ISO Standard 7730, 56

janitors, 37–38
Jennings, Eric, 11, 241n52
Johnson, Warren S., 37–38
Johnson Electric Service Company, 38; Furnace Draft Regulator, 40, *41*, 44
Judgment of the Storm (1924), 82

Keaton, Buster, 82
Kember, Sarah, 223
Kenya, 169, 182–83, *184*
Kiesling, Barrett C., 200
Kipling, Rudyard, 79
knowledge, 120, 222, 241n52; scientific, 42, 75, 130, 154; thermal, 11, 19, 59, 75, 185, 196
Kowal, Emma, 112, 195

labor, 7, 38, 122; productivity, 31–32, 64–65; racialized, 91–92; thermal exposure and, 13, 36–37

Langley, Samuel Pierpont, 171
lasers, 156–57, 159
Lévi-Strauss, Claude, 10
Lidar systems, 159, 161
life and nonlife, 7, 11, 187
light waves, 137, 144–45, 156
lithographers, 197, 198
Lovecraft, H. P., 96–97, *98*

MacMillan, Donald B., 80, 92–93
magnetic media, 206–7
Maillefert, Arthur, 119, 122
Marks, Laura, 77, 167
Martin, David, 122
Martin, Niall, 187
Marx, Karl, 220
mass media, 19–20, 79, 137, 154, 203, 217;
    failures, 145, 146; infrared technologies,
    138, 155, 168; modern media as, 194;
    thermal attractions and, 95, 104, 108;
    transitions, 70, 221
materiality, 20, 83, 115, 222; media's, 15, 24,
    154, 194; of temperature, 9, 168
Mbembe, Achille, 7
McLuhan, Marshall, 9, 14, 217, 222–23
McPherson, Tara, 191
media studies, 4, 11, 18, 223
medical practice, 6, 11, 110, 121, 137; elec-
    trotherapy, 151; infrared imaging, 26,
    168–69, 173–74, *174*, 177, 179, 184; light
    therapy, 148; radio heat and, 150–51, 153
meltings, 210, 219–22, 224
metallurgy/metallurgical approach, 20, 25,
    113, 222
Mexican immigrants, 110–11
military, 104, 120, 183–84; Active Denial
    System (heat ray), 1, 17, 135–39, 153, 155;
    infrared imaging, 170, 185–88
mines, 162, 203; Morenci (AZ), 179, *180*
Minneapolis-Honeywell Regulator Com-
    pany, 34, 66, 212, 217; advertisements,
    40, 42, *43*, *44*; energy conservation, 50;
    marketing of "discomfort," 45, 102; smart
    thermostats, 61–62; thermostat design,
    46, *47*; Vector Occupant app, 64; zone
    heating campaign, 45, 48
Mississippi State Penitentiary, 125, 130
mobile technology, 62, 65, 182

modernity, 11, 33
Mosse, Richard, 170, 185–87, *186*
Mr. Robot (2015), 209–10, 218
Mukherjee, Rahul, 16, 131, 136
Mumford, Lewis, 10

Nading, Alex, 226n24
Nanook of the North (1922), 81–83, 86–88,
    87, *90*
National Broadcasting Company (NBC),
    203, *204*
National Energy Act (1978), 176
neoliberalism, 5, 27, 35, 69, 70
Nest thermostats, 35, 61–63, 66–69, 142
neutrality: carbon, 211; thermal, 58, 73, 107,
    170, 185, 187
New York State Commission on Ventila-
    tion, 53
New York Times, 32, 51, 61, 208
Nietzsche, Friedrich, 250n12
Night, Darryl, 109
Nippon Telephone and Telegraph, 158
nitrate film, 194, 206
Nixon, Richard, 50
Nixon, Rob, 112
Noctovisor, 147–48, 155
Noon, Derek, 192
normativity: bodily, 188, 190; temperature,
    11, 36, 69, 88, 170, 176; thermal, 182, 185,
    188

objectivity. *See* thermal objectivity
oil, 50, 58, 162; heating systems, 38, 39
Ong, Boon Lay, 8, 18, 73
oppression, 113–14
optical communications, 144, 155–58. *See
    also* fiber-optic cables
optimal temperature, 31–32, 53, 55, 56, 209;
    ideal cooling conditions, 100–101, 129;
    neutrality, 72–73, 107
Osagboro, Muti-Ajamu, 127
Osman, Michael, 38–39, 41, 226n17, 229n13,
    230n33
Otherness, 77, 91, 92, 95, 96–97, 104
overheating, 102, 113, 199, 206; of houses,
    42; of prisons, 1, 22, 124–28; of schools,
    129, 243n83; of voting machines, 19. *See
    also* sweatboxing

paper production, 197–99, 208, 250n26
Parks, Lisa, 169, 187
Parr, Joy, 128
Parsons, Ken, 59
participatory thermal sensing, 64–66
patents, 62, 205
Peguero, Raymond, 126
Peltier effect, 105–6
permafrost, 210, 219
personalized technologies, 5, 35–36, 131–32;
    cooling products, 22, 77–78, 104–7; radi-
    ant thermal media, 154–55; thermostat
    control, 61–69, 70
Peters, John Durham, 223
Peterson, Marina, 18, 128
phase transitions, 220–21
Philco Radio and Television Corporation, 103
Phillips, Lee, 128
phonograph, 24, 91, 94; cylinders, 194,
    205–6, 208, 217
photoelectric cells, 146, 148
photography, 9–10, 104; infrared, 148, 168,
    171–73, 172, 175, 177, 188; negatives, 206
photophone, 23, 25, 137, 144–46, 155–56, 164
Picture Theater Advertising (Sargent), 89,
    92, 95
plantations, 113, 116, 118, 121, 127
polar explorations, 75, 80, 81, 95; films on,
    25, 81–83, 103; Lovecraft's stories on,
    96–97, 98; radio shows on, 92–93, 103
police: interrogations, 120; violence, 51, 109,
    111–12
Popular Mechanics, 146, 172
Popular Science Monthly, 146, 176–77
Povinelli, Elizabeth, 7
power. *See* thermopower
power industry, 162, 175
preservation of media, 24, 194–95, 206–7,
    208
print industry, 194, 197–99, 208
prisons, 28, 136, 222; chain gang films, 119;
    death row, 122, 124–25; improvements
    to conditions in, 128, 131; overheating
    of, 1, 22, 113, 124–27; use of sweatboxes,
    117–19, 122, 123, 241n34, 241n38
progress, 25, 195–96, 210

quantum computers, 191–92, 193, 194, 208

race and racialization, 27, 56, 115; coldness
    and, 75, 91; depicted in fiction, 97, 98;
    heat images and, 170, 172–73, 185–88;
    labor and, 13, 38, 45, 91–92, 227n48;
    sweatboxing and, 113–14, 116, 118–121,
    241n34; thermal vision and, 184–85
racism, 12, 87, 92, 104; environmental
    determinists and, 75, 195; medical, 121;
    treatment of prisoners and, 118–19, 127
radiant emissions. *See* emissivity
radiant heating, 16, 31, 39, 130, 143, 154; ex-
    perimentation and technologies, 137, 142
radiant infrastructures, 16, 136
radiant thermal media, 141, 142, 145,
    153–55, 165; description of, 15–16, 136–37;
    examples of, 23; fiber-optic cables as, 164;
    infrared camera as, 168. *See also* infrared
    cameras and imaging
Radin, Joanna, 112, 195
radio: air-conditioning and, 102–3, 202–3,
    204; "fever," 137, 149–51; frequency cur-
    rents, 151; as a medium, 154; shows, 21,
    80, 92–94, 94, 103–4; stations, 78; trans-
    missions, 16, 79, 105, 149, 154, 165
radiographic episteme, 169, 189
radiotherm, 150, 151, 153
radio waves, 16, 23, 143, 145; heat effects of
    shortwave, 3, 137, 149, 151–54, 155, 164–65;
    light waves *vs.*, 156; long-wavelength,
    139, 153
Raiford Prison. *See* Union Correctional
    Institution
Rankine, Oliver, 145–46, 149
Rawling, S. O., 172
Rayleigh scattering, 157, 158, 163
refrigeration/refrigerators, 74, 97, 100–101,
    194, 195
refugees, 170, 185–87
reindeer, 219–20
Rejali, Darius, 120
remote control, 48, 62, 66–67, 153, 155
Ridley, Henry, 118–19
rights. *See* thermal rights
Roctest, 162
Rubio, Fernando Domínguez, 8, 194, 207

Sackett and Wilhelms, 199, 200
Sargent, Epes, 89, 92, 95

Satellite Imaging Corporation, 179, *180*

Schüll, Natasha, 63

Scientific American, 155, 200

Scientific Diathermy Corporation, 153

Scott, Janie, 116

Seattle (WA), 210–11

SecuSystems, 184

Seiko Thermic, 105

selenium cell, 144–46, 157

sensing practices, 168

sexual harassment, 51

Shachtman, Tom, 73

Shakur, Assata, 127

Sharpe, Christina, 129

ships, 117–18, 127–28, 187, 240n13

silicon cell, 157

Simmons, Kristen, 87

Singapore, 213–14, 216

slavery, 22, 113, 116–17, 121–22, 129; ships,
127–28

Sloterdijk, Peter, 79

SMART (science monitoring and reliable
telecommunications), 163

Smith, Jen Rose, 11, 75, 88

Smith, Prince, 116

snow films: colonization and racialization
and, 87–88; exhibitors and marketing
of, 88–91, *90*; exploitations, 86–87, *87*,
89; popularity of, 21, 80–83, 103; real *vs.*
artificial snow in, 83–84, *85*

solidifications, 220–21, 224

Souls for Sale (1923), 84

soundproofing, 202–3

sound recording, 204–5

South Africa, 169, 182, 184–85, 203

Southern California Edison, 67–68

spatiotemporality, 10, 14, 17, 18, 129; of
heat, 138; of ice, 76

spectroscopy, 26, 140, 145

standardization, 21, 34, 197, 206, 208, 209;
of bodies, 6, 17; cinema's, 199, 202; of
temperature, 52–53, 101; of thermal
comfort, 54, *55*, 56–60, 101; thermopower
and, 69–70

Standing Rock (ND), 111

Stonechild, Neil, 109

studio productions, 24, 201, 202–3

suicides, 126–27

summer, 32, 42, 50, 65, 107, 153; moviegoing
in, 88–89, *90*, 92, 99, 103; prison conditions in, 122, 124, 127; sweatboxes and, 22,
115–16, 120

sun, 17, 115–16, 121, 155, 171; agriculture
and, 181; solar technology, 144, 216; thermal radiation, 15, 73, 130, 142, 143, 164

surveillance, 184, 187; fiber-optic cables
and, 162, 165; infrared imaging and, 24,
26, 170. *See also* drones

sweatboxing, 22, 25, 113–14; history and
racialization of, 115–21, 127–28, 129, 131;
practices in prisons, 122, *123*, 124–27

Sweatt, Harold W., 46

Taggart, Jonathan, 9, 17

Tainter, Charles Sumner, 23, 144

Tawil-Souri, Helga, 139, 165

telecommunications cables, 163–64

telegraph, 16, 78, 154, 158; heat ray (infra-red),
17, 23, 24, 138, 146, *147*, 164; wireless, 79, 145

telephone, 9, 16, 78, 144, 146, 154, 217

television, 9, 23, 194, 217; infrared rays and,
23–24, 137, 146–49, 155; signals, 143, 155

temperature: air-conditioning, 100–102;
body, 2, 16, 18, 169; computing, 191–92;
evenness or stabilization of, 39, 42, 45,
203, 204–7, 208, 210; extremes, 109–11,
122, 124–27, 128, 205, 219; fluctuations, 4,
24, 42–43, 62, 205–7; link to civilization,
10–11, 25; materiality of, 9, 168; mean
annual, 195, *196*, 215, 216; meanings and
functions, 1–3; patterns of existence and,
18–19; perceptions of, 166–68, 189–90;
power of, 27; radiant emissions and,
139–42, 176, 247n27; as representation,
39, 45; sensing of fiber-optic technology, 161–64; separation from energy, 34,
37, 45, 71; visible light and, 143. *See also*
climate(s); optimal temperature

temperature control, 6, 14, 17, 21–22, 33,
96, 132; aesthetic of, 46; family bonding and, 40; film manufacturing and,
199–202; gender conflicts, 32, 35, 48–52,
49, 167–68; infrastructures of, 24; origins
of, 38; personalized, 61–69, 70, 77, 106,
131; phonograph cylinders and, 205–6;
printing industry and, 197–199, 250n26;

in prisons, 124–27; regulations, 50–51; standardization, 52–53, 59; storage environments, 206–7; in studios, 202–3; in theaters, 102. *See also* thermostats

temporality. *See* spatiotemporality

Tesla, Nikola, 151

thallium sulfide cell, 145

thermal autonomy, 28, 35, 66, 70–71, 77, 224; politics of, 23, 115; thermal violence and, 130–32

thermal capitalism, 13, 31–32, 64, 68, 108, 221; cold technologies and, 74, 106; thermal attractions and, 73, 77; undoing of, 209, 218

thermal comfort, 21, 32, 45, 189; contrast and, 101–2; industrial, 64–65; participatory, 64; personalization of, 35, 61–69, 106–7; in prisons, 124–25; researchers and studies, 35, 57–60, 107; standards, 34, 56–57, 101, 113, 124; thermal desires and, 72–73. *See also* comfort zone chart

thermal communications, 2–3, 13, 34, 65, 77, 126; atmospheric forms of, 15; attributes of, 16–17; body as a receiver of, 151, 223; environment as a medium for, 190; media studies and, 4–5, 225n5; radiant, 138; thermoception and, 35, 56; transformation of, 70, 136

thermal desires, 20–21, 34, 45, 63; autonomy and, 66, 77; experiences of hot and cold and, 72–74, 101

thermal difference, 7, 21, 32–35, 48, 52, 69; acknowledgment of, 46, 57, 58, 70, 106; agriculture and, 181; of bodies, 187–88; of inside and outside, 166; normal/abnormal temperature and, 170, 176; reconciliation of, 66

thermal entanglement, 27, 131, 157, 164, 203, 223–24; of computing infrastructures, 192, 194, 213, 214; description of, 18, 166; of digital systems, 208; of the phonograph player, 205; of the print industry, 197–98

thermal exposure, 2, 12, 13–14, 68, 77, 112, 115; audiovisual exposure and, 99

thermal media, 1, 6–9, 24–27, 70, 108, 114, 222; coldsploitation and, 76–77, 94, 103; genealogies, 19, 20; genres, 21; goal of, 107; infrared cameras as, 168–69, 177; key attributes, 16–19; local, 28, 224; politics

of, 223; primary forms of, 15–16; spatiotemporal practices, 14. *See also* radiant thermal media

thermal objectivity, 2, 3, 7, 8, 22, 222–23; of infrared images, 170, 188, 190

thermal rights, 2, 22, 128–29, 223–24

thermal violence, 7, 8, 17, 28, 87, 142, 154, 223–24; climate change and, 22, 27, 222; of heat ray guns, 135–36; invisible nature of, 126; paradigm of exposure, 129–30; in prisons, 122, 124–28; sweatboxing as, 22, 113–14, 115–21; thermal autonomy and, 130–32; thermostat control and, 51–52, 69; weaponizing the environment and, 109–12

thermal vision, 24, 168; description, 166–67; difference and, 187–88; haptic visuality and, 21, 76–77, 90; machinic, 181; militarized, 183–84; political potential of, 189–90; *Predator* vision, 170, 185, 188

thermoception, 4–6, 17, 31, 52, 171, 181, 223; as anticipatory, 127; of comfort, 56, 57; Goldilocks theory of, 72; regimes, 7, 24, 114, 169, 170, 175, 177, 179; standards and, 34–35, 60; thermal vision and, 167, 189–90, 247n5; vernacular forms of, 48, 52, 222

thermocultures, 5–8, 17, 19–20, 45, 222, 224; climate change and, 27; cool attractions and, 106, 107; infrared camera and, 188; maintenance of media and, 204, 207; race and, 13, 122; shift in, 35, 94; violence and, 113, 131; visual indicators of, 166, 168

thermodynamics, 2, 9–10, 16, 72

thermoelectric technology, 105–6

thermography, 169, 172–73, 175–77, 248n28

thermometer, xiv, 2, 14, 37, 95; fiber-optic cable as, 161–64; global internet as, 24, 139; medical practice and, 6, 7

TherMOOstat, 64

thermopolitics, 7, 27

thermopower, 2–3, 12, 33, 35–36, 207, 212, 217; body temperature and, 114, 170; climate change and, 216, 223; networked, 69–71; spectral negotiations of, 136–37, 138–39, 142, 153, 165; stabilization of temperature and, 194, 196, 203, 206, 208; thermal attractions and, 73, 77, 94, 104; uneven distribution and, 7; vectors of, 168, 190 thermostats, xiv, 6, 20–21, 23, 28, 207, 217; automated and programmable, 42–43, 43; capitalism and, 209, 218; control and gender conflicts, 32, 35, 43–44, 44, 48–52, 49, 167; design, 46, 47; early technologies, 37–40, 41, 229n11, 229n13, 229n21; market, 45–46; office settings, 31–32; setpoint, 52, 58, 100, 167; smart or AI-enabled, 21, 26, 61–69, 106, 212; social origins, 34, 36–37

Thompson, Darla, 116

tourism, 11, 21, 104, 184

trains, 102–3

Tredgold, Thomas, 39

Tuck, Eve, 113–14

Typhoon Fan Company, 95, 96

ultraviolet light, 15, 144, 160

undersea cables, 163, 164, 213

Union Correctional Institution, 122, 123, 124, 125, 130

Universal Film Manufacturing, 200

universality, 2, 13, 74; standard of comfort, 21, 35, 54, 56–57, 59, 115; thermal, 107, 114, 115

US Customs and Border Protection (CBP), 110

utility companies, 35, 63, 66–69, 71, 131

vacuum tubes, 149–50, 151

Vannini, Phillip, 9, 17

Vassallo, Susi, 130

Vaughan, Hunter, 84

Velkova, Julia, 197, 210

ventilation systems, 34, 79, 95–96, 100; in prisons, 124; standardization of, 52–53

visible light, 15, 138, 153; blinding effects of, 155, 160, 165; early experiments with, 146, 147, 149; fiber optics and, 160, 160–61; infrared cameras and, 168, 171, 175–76; lasers as, 157; spectrum of, 16, 140

voting machines, 19, 227n62

Walker, Janet, 220

Washington, Harriet, 121

waste-heat recovery, 194, 210–11

water cannons, 2, 111, 112, 129, 142

water protectors (Standing Rock), 111

Watt-Cloutier, Sheila, 27–28

wax cylinders, 204–5, 208

Way Down East (1920), 83

weaponizing the environment, 22, 110–13, 121, 126, 131–32, 136

weather: cold waves, 79–80; film production and, 84; heat waves, 42, 68, 92, 124, 128; infrared transmission and, 155, 156, 160, 165; mechanical, 200; phonograph cylinders and, 205–6; rain, 219–20; viral transmissions and, 26, 219. See also humidity; wind

Well, H. G., 136, 145

Wernimont, Jacqueline, 228n63

Westin Building Exchange, 210–11

Westinghouse, 86, 105, 152

Westworld (2016), 209–10

whiteness, 75, 88, 92, 95, 185, 223

white prisoners, 118–19, 122, 241n34

white supremacy, 75, 125, 127, 195, 196

Whitney, Willis R., 149–50, 152

Whyte, Kyle Powys, 22, 222

wildlife management, 169, 182–85, 183

Willat Film Manufacturing Corporation, 200

wind, 93, 97; machines, 83–84

windows, 4, 17, 31, 48, 53, 110, 201

winter, 79–80, 109, 111, 154, 198, 201; prison conditions in, 122; sweatboxes and, 115–16. See also snow films

Wishart, Paul, 217

women, 37, 39, 110; in prison, 125, 127; as targets of advertising, 40–42; thermostat control and, 32, 44, 48–51, 167–68

Wood, Robert Williams, 171, 175, 188

World War I, 19, 90, 120, 138, 145, 155; cold weather and, 79–80, 201

World Wildlife Fund (WWF), 182–84, 183

wristbands, 15, 22, 106–7, 142

Yamal Peninsula, 219–20

Yusoff, Kathryn, 22, 222

Zenith Radio Company, 92, 94

zone heating, 46, 48, 231n60

Zylinska, Joanna, 223